T0257912

Petrology: Innovative Approaches and Applications

Petrology: Innovative Approaches and Applications

Edited by **John Wayne**

New York

Published by Callisto Reference,
106 Park Avenue, Suite 200,
New York, NY 10016, USA
www.callistoreference.com

Petrology: Innovative Approaches and Applications
Edited by John Wayne

International Standard Book Number: 978-1-63239-509-2 (Hardback)

Printed in the United States of America.

Contents

Permissions

List of Contributors

Preface

Petrology is the branch of geology that deals with the origin, structure, and composition of rocks. This book has been compiled for professionals as well as for advanced graduate courses in petrology. It consists of research-focused information about the current developments and applications of modern petrologic and geochemical methods for comprehending metamorphic, igneous and even sedimentary rocks. Research analyses compiled in this book provide an overview on the applications of latest petrologic methodologies to rocks of diverse origins. They represent a broad variety of settings (from Africa to Central Asia, and from South America to the Far East) as well as ages ranging from late Precambrian to late Cenozoic, and also Mesozoic/Cenozoic volcanism.

This book is the end result of constructive efforts and intensive research done by experts in this field. The aim of this book is to enlighten the readers with recent information in this area of research. The information provided in this profound book would serve as a valuable reference to students and researchers in this field.

At the end, I would like to thank all the authors for devoting their precious time and providing their valuable contribution to this book. I would also like to express my gratitude to my fellow colleagues who encouraged me throughout the process.

Editor

Secular Evolution of Lithospheric Mantle Beneath the Central North China Craton: Implication from Basaltic Rocks and Their Xenoliths

Yan-Jie Tang, Hong-Fu Zhang and Ji-Feng Ying
State Key Laboratory of Lithospheric Evolution, Institute of Geology and Geophysics,
Chinese Academy of Sciences, Beijing
China

1. Introduction

The old lithospheric mantle beneath the North China Craton (NCC, Fig. 1a) was extensively thinned during the Phanerozoic, especially in the Mesozoic and Cenozoic, resulting in the loss of more than 100 km of the rigid lithosphere (Menzies et al., 1993; Fan et al., 2000). This inference comes from the studies on the Ordovician diamondiferous kimberlites (Fig. 1b), Mesozoic lamprophyre-basalts and Cenozoic basalts, and their deep-seated xenoliths (e.g. Lu et al., 1995; Griffin et al., 1998; Menzies & Xu, 1998; Zhang et al., 2002). This remarkable evolution of the subcontinental lithosphere mantle, which has had profound effects on the tectonics and magmatism of this region, has attracted considerable attention (e.g. Guo et al., 2003; Deng et al., 2004; Gao et al., 2004; Rudnick et al., 2004; Xu et al., 2004; Ying et al., 2004; Zhang et al., 2004a, 2005, 2008; Wu et al., 2005; Tang et al., 2006, 2007, 2008, 2011; Zhao et al., 2010). However, the cause of such a dramatic change, from a Paleozoic cold and thick (up to 200 km) cratonic mantle (Griffin et al., 1992; Menzies et al., 1993) to a Cenozoic hot and thin (< 80 km) "oceanic-type" lithospheric mantle, is still controversial.

Based on the Mesozoic basalt development, Menzies and Xu (1998) argued that thermal and chemical erosion of the lithosphere was perhaps triggered by circum-craton subduction and subsequent passive continental extension. This suggestion was first supported by the geochemical studies on the Mesozoic basalts and high-Mg# basaltic andesites on the NCC (Zhang et al., 2002, 2003). A partial replacement model was proposed, having a sub-continental lithospheric mantle in this region composed of old lithosphere in the uppermost part and newly created lithosphere in the lower part (Fan et al., 2000; Xu, 2001; Zheng et al., 2001). The clearly zoned mantle xenocrysts found in Mesozoic Fangcheng basalts (Zhang et al. 2004b) provide the evidence for such a replacement of lithospheric mantle from high-Mg peridotites to low-Mg peridotites through peridotite-melt reactions (Zhang, 2005). Another different model was also proposed that ancient lithospheric mantle was totally replaced by juvenile material in the Late Mesozoic (Gao et al., 2002; Wu et al., 2003). On the basis of Os isotopic evidence from mantle xenoliths enclosed in Cenozoic basalts, Gao et al. (2002) suggested that two times replacement existed in the NCC. They attributed the replacement of the old lithospheric mantle beneath the Hannuoba region to the collision of the Eastern

Block with the Western Block and the second time perhaps to the collision of the Yangtze Craton with the NCC. Based on the study of Mesozoic Fangcheng basalts, Zhang et al. (2002) proposed that the replacement of the lithospheric mantle beneath the southern margin of the NCC was triggered by the collision between the Yangtze and the NCC. Zhang et al. (2003) further suggested that the secular lithospheric evolution was related to the subduction processes surrounding the NCC, which produced the highly heterogeneous Mesozoic lithospheric mantle underneath the NCC (Zhang et al., 2004a). In contrast, Wu et al. (2003) thought that subduction of the Pacific plate during the Mesozoic was the main cause of lithospheric thinning. Meanwhile, Wilde et al. (2003) correlated this event with the lithospheric thinning resulting from the breakup and dispersal of Gondwanaland and suggested that the removal was partial loss of mantle lithosphere, accompanied by wholesale rising of asthenospheric mantle beneath eastern China.

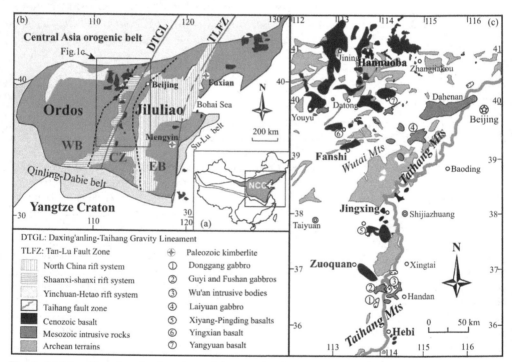

Fig. 1. (a) Map showing the location of the North China Craton (NCC); (b) Three subdivision of the NCC (modified from Zhao et al., 2001). Two dashed lines outline the Central Zone (CZ), the Western Block (WB) and the Eastern Block (EB); (c) The distribution of Cenozoic basalts, Mesozoic mafic intrusive rocks and of Archean terrains in the studied area.

Based on the Daxing'anling-Taihang gravity lineament (DTGL), the NCC can be divided into western and eastern parts (Ordos and Jiluliao terrains, Fig. 1b). The temporal variations in geochemistry of Cenozoic basalts from both sides of the DTGL suggest an opposite trend of lithospheric evolution between the western and eastern NCC (Xu et al., 2004), i.e. the progressive lithospheric thinning in the western NCC and the lithospheric thickening in the eastern NCC during the Cenozoic. Considering that the Taihang Mountains are in the Central Zone of the NCC, which geographically coincides with the DTGL (Fig. 1b), the

Secular Evolution of Lithospheric Mantle Beneath the Central North China Craton: Implication
from Basaltic Rocks and Their Xenoliths

3

Mesozoic-Cenozoic lithospheric evolution beneath this region is an important issue to comprehensively decipher the mechanism for the lithospheric evolution beneath the NCC. In this paper, a summary of geochemical compositions of Mesozoic gabbros, Cenozoic basalts and their peridotite xenoliths in the Central Zone are presented to trace the petrogenesis of these rocks, the Mesozoic-Cenozoic basaltic magmatism, and further to discuss the potential mechanism of the lithospheric evolution in this region.

2. Geological background and petrology

The NCC is one of the oldest continental cratons on earth (3.8~2.5 Ga; Liu et al., 1992a) and is composed of two Archean nuclei of Eastern and Western Blocks (Fig. 1b). The Eastern Block has thin crust (<35 km), weakly negative to positive Bouguer gravity anomalies and high heat flow because of widespread lithospheric extension during Late Mesozoic and Cenozoic, which produced the NNE-trending North China rift system (Fig. 1b), and the lithosphere is inferred to be <80~100 km (Ma, 1989). The Western Block has thick crust (>40 km), strong negative Bouguer gravity anomalies, low heat flow and a thick lithosphere (>100 km) (Ma, 1989). The Yinchuan-Hetao and Shanxi-Shaanxi rift systems (Fig. 1b) appeared in the Early Oligocene or Late Eocene, and the major extension developed later in the Neogene and Quaternary (Ye et al., 1987; Ren et al., 2002).

The basement of the NCC is composed of amphibolite to granulite facies rocks, such as Archaean grey tonalitic gneisses and greenstones and Paleoproterozoic khondalites and interlayered clastic, and an overlying neritic marine sedimentary cover (Zhao et al., 1999, 2001). It was considered that the NCC underwent the ~1.8 Ga subduction/collision between the Eastern and Western Blocks (Zhao et al., 1999, 2001) resulting in the amalgamation of the NCC. The east edge of the orogenic belt coincides with the Taihang Mountains rift zone.

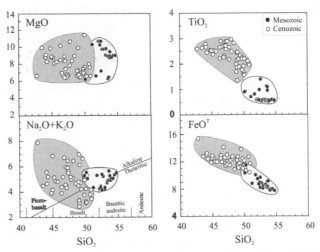

Fig. 2. Major oxide variations of the Mesozoic and Cenozoic basaltic rocks from the Central Zone. Data sources: Cenozoic basalts (Zhou & Armstrong, 1982; Xu et al., 2004; Tang et al., 2006), Mesozoic rocks (Cai et al., 2003; Chen et al., 2003, 2004; Chen & Zhai, 2003; Peng et al., 2004; Zhang et al., 2004), classification of volcanic rocks (TAS diagram, Le Bas et al., 1986), the boundary between alkaline and tholeiitic basalts (Irving & Baragar, 1971).

In the Central Zone of the NCC, the Mesozoic mafic intrusions are widespread, e.g. Donggang, Guyi, Fushan gabbros (150~160 Ma), Wuan monzonitic-diorites (126~127 Ma), Laiyuan gabbro, Wang'anzhen and Dahenan monzonites (135~145 Ma) (Fig. 1c), which were cut by minor, late stage calc-alkaline lamprophyres (~120 Ma) that occur as dykes or small intrusions (Chen et al., 2003, 2004; Chen & Zhai, 2003; Peng et al., 2004 and references therein; Zhang et al., 2004a). These Mesozoic gabbros are of small volume and occur as laccoliths, knobs, or as xenoliths in Mesozoic dioritic intrusions.

Cenozoic basalts in the Central Zone (Fig. 1c) are distributed in the Hebi (~4 Ma), Zuoquan (~5.6 Ma), Xiyang-Pingding (7~8 Ma) and Fanshi-Yingxian regions (24~26 Ma) (Liu et al., 1992b), which are mainly composed of alkaline basalts and olivine basalts, including alkaline and tholeiitic sequences (Fig. 2). Abundant mantle-derived peridotite xenoliths are found in the basalts from the Fanshi and Hebi regions (Zheng et al., 2001; Xu et al., 2004), and mantle olivine xenocrysts are entrained in the Xiyang-Pingding basalts, which are interpreted as the relict of old lithospheric mantle (Tang et al., 2004).

3. Methodology and samples

Experiments have demonstrated that more SiO_2-undersaturated magmas are produced at higher pressures than tholeiitic lavas (e.g., Falloon et al., 1988). Because the lithospheric mantle and asthenosphere generally are different in geochemical signatures, it can be inferred that the lithosphere is >80 km thick if the alkali basalts have an isotopic signature of sub-continental lithospheric mantle. Conversely, if the tholeiitic basalts have an asthenospheric signature the lithosphere is inferred to be <60 km thick (DePaolo and Daley, 2000). The geochemistry of mantle-derived magmas is dependent on the depth of melting (Herzberg, 2006), thus the geochemistry of basaltic rocks can be used to monitor variation in lithospheric thickness and geochemistry through time (e.g., DePaolo and Daley, 2000).

Ideally, tracing the chemical evolution of the mantle lithosphere would be accomplished by measuring the compositions of coherent, pristine suites of direct mantle samples, lacking metasomatic overprints, and with a well-determined age and geological context. The chemical compositions of direct mantle samples such as abyssal peridotites and peridotite xenoliths, and of indirect probes of the mantle such as basalts from MORBs and OIBs, have provided strong evidence for chemical complexity and heterogeneity of the mantle (Hofmann, 2003). Complexity in the interpretation of chemical compositions of basalts often results from the modification of primary melt compositions due to crustal contamination during their generation and ascent. For this reason, the most primitive basalts, usually with the highest-MgO content, are taken to be the least affected by crustal interaction and therefore the best record of mantle compositions.

Mesozoic basaltic rocks in the Central Zone are dominantly gabbroic intrusions, which are derived from lithospheric mantle (Tan & Lin, 1994; Zhang et al., 2004). Some of them contain peridotite and/or pyroxenite xenoliths (Xu & Lin, 1991; Dong et al., 2003). Previous petrological and geochemical studies indicate that the gabbroic rocks have compositions of original basaltic magmas (Tan & Lin, 1994; Zhang et al., 2004). Although some workers report crustal contamination (Chen et al., 2003; Chen & Zhai, 2003; Chen et al., 2004), others suggest that in many cases isotopic composition of these rocks still reflect variation in the mantle source and can provide the information on the continental lithospheric mantle beneath the region (Tan & Lin, 1994; Dong et al,. 2003; Zhang et al., 2004).

Secular Evolution of Lithospheric Mantle Beneath the Central North China Craton: Implication
from Basaltic Rocks and Their Xenoliths

5

In contrast, the geochemical features of Cenozoic basalts from Taihang Mountains (Tang et al., 2006), are very similar to those of the Cenozoic Hannuoba basalts (e.g. Zhou & Armstrong, 1982; Song et al., 1990; Basu et al., 1991), suggest their derivation mainly from asthenosphere with negligible crustal contamination. The occurrence of mantle xenoliths and xenocrysts suggests that these lavas ascended rapidly, implying that significant interaction with crustal wall rocks could not happen. So, their chemical compositions can be used to probe their mantle sources. Although these basalts are dominantly of asthenospheric source, their variable Sr-Nd isotopic ratios indicate some contributions of lithospheric mantle (Tang et al., 2006), whereby we could indirectly trace the feature of the Cenozoic mantle lithosphere. Meanwhile, some available data of mantle xenoliths entrained in these Cenozoic basalts can be used to directly infer the nature of the lithospheric mantle beneath the craton.

Due to the biases brought about by variable assimilation-fractional crystallization processes, we use only gabbros and basalts with the geochemical compositions of relatively primitive samples (MgO >6 wt.%) from each region, as well as their hosted peridotite xenoliths, to study the nature of mantle lithosphere beneath the Central Zone of the NCC.

4. Variations in geochemical compositions

Figures 2-7 show clear variations in geochemical compositions between the Mesozoic and Cenozoic basaltic rocks in the Central Zone. Compared with the Cenozoic basalts, the Mesozoic mafic intrusive rocks are: (1) higher in SiO_2, lower in FeO^T and TiO_2 contents (Fig. 2); (2) enriched in light rare earth element (LREE) and large ion lithophile element (LILE, such as Ba, Th and U), but depleted in high field strength element (HFSE, e.g. Nb, Ta, Zr and Ti; Figs. 3 & 4); (3) high Sr and low Nd and Pb isotopic ratios (most $^{87}Sr/^{86}Sr_i=0.705\sim0.7065$, $^{143}Nd/^{144}Nd_i<0.512$; Fig. 5; $^{206}Pb/^{204}Pb_i<17.5$, $^{207}Pb/^{204}Pb_i<15.5$, $^{208}Pb/^{204}Pb_i<38.0$, Fig. 6), typically EM1 features. These features are completely different from those of MORB, OIB and Cenozoic basalts in this region, which are generally lower in SiO_2, higher in FeO^T and TiO_2 contents (Fig. 2), depleted in Sr-Nd isotopes (Fig. 5) and have no HFSE depletion (Figs. 3 & 4). These geochemical distinctions reflect their mantle source differences between Mesozoic and Cenozoic times.

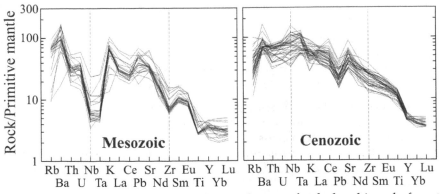

Fig. 3. Primitive mantle-normalized trace element diagrams for the basaltic rocks from the Central Zone. Data sources: primitive mantle (McDonough & Sun, 1995), others as in Fig. 2.

5. Discussion

5.1 Petrogenesis of Cenozoic basalts and lithospheric thickness

Cenozoic basalts from the Taihang Mountains have many similar features to those of Cenozoic Hannuoba basalts (Zhou & Armstrong, 1982; Peng et al., 1986; Song et al., 1990; Basu et al., 1991; Liu et al., 1994) and many alkali basalts from both oceanic and continental settings (Barry & Kent, 1998; Tu et al., 1991; Turner & Hawkesworth, 1995) in their elemental and isotopic compositions (Figs. 2-7). Their common geochemical features of OIB and/or MORB are interpreted as having been derived from the asthenospheric mantle.

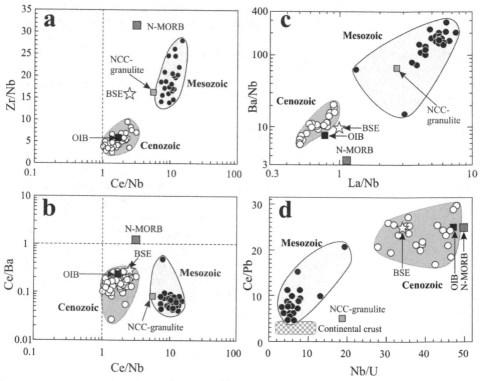

Fig. 4. Variations in trace-element ratios for the basaltic rocks from the Central Zone. Data sources: BSE, N-MORB and OIB (Sun & McDonough, 1989; McDonough & Sun, 1995); NCC-granulite, the average composition of old granulite terrains on the NCC (Gao et al., 1998); Continental crust (Rudnick & Gao, 2003). Other data sources and symbols as in Fig. 2.

Their incompatible trace element ratios, e.g. Ba/Nb, La/Nb, Zr/Nb, Ce/Nb, Ce/Ba, Nb/U and Ce/Pb values, are very close to those of OIB (Fig. 4). Some slightly lower and variable Nb/U ratios for these Cenozoic basalts (Fig. 4d) might suggest the involvement of lithospheric mantle in their source, because the metasomatised lithospheric mantle is probably involved in producing the negative Nb anomalies (Arndt & Christensen, 1992). Moreover, the lower initial ratios of $^{143}Nd/^{144}Nd_i$ (<0.5125) and higher $^{87}Sr/^{86}Sr_i$ (>0.705; Fig. 5) also indicate the involvement of old lithospheric mantle beneath the NCC. Three low ratios of Pb isotopes ($^{206}Pb/^{204}Pb_i$<16.9) of the Cenozoic basalts (Fig. 6) are close to the field

Secular Evolution of Lithospheric Mantle Beneath the Central North China Craton: Implication
from Basaltic Rocks and Their Xenoliths

7

of the Smoky Butte lamproites that were believed to have been derived from ancient EM1-type lithospheric mantle (Fraser et al., 1985). They are also similar to those of Cenozoic potassic basalts in the Wudalianchi, northeastern China (Zhang et al., 1998), whose source is interpreted as metasomatically enriched mantle. Integrating the isotopic ratios with the element compositions, the Cenozoic basalts from the Taihang Mountains are inferred to be derived from partial melting of an asthenospheric source with different degrees of the involvement of old lithospheric mantle.

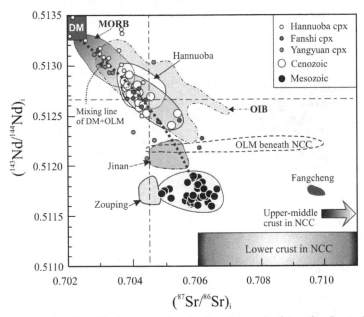

Fig. 5. ^{87}Sr/^{86}Sr$_i$ vs. ^{143}Nd/^{144}Nd$_i$ diagrams for the basaltic rocks from the Central Zone, compared with the Hannuoba basalts (Song et al., 1990; Zhi et al., 1990; Basu et al., 1991; Xie & Wang, 1992), old lithospheric mantle (OLM) beneath the NCC (Zhang et al., 2002), CPX in peridotite xenoliths in the Fanshi (Tang et al., 2008; 2011), Yangyuan (Ma & Xu, 2006) and Hannuoba basalts (Song & Frey, 1989; Tatsumoto et al., 1992; Fan et al., 2000; Rudnick et al., 2004), DM, MORB and OIB (Zindler & Hart, 1986), Mesozoic Fangcheng basalts (Zhang et al., 2002), Mesozoic Jinan gabbros (Zhang et al., 2004) and Zouping gabbros (Guo et al., 2003; Ying et al., 2005), the upper-middle crust and lower crust of the NCC (Jahn & Zhang, 1984; Jahn et al., 1988). Other data sources and symbols as in Fig. 2.

The clinopyroxenes (CPX) in mantle peridotite xenoliths entrained in the Cenozoic basalts have significant variations in Sr-Nd isotopic compositions (^{87}Sr/^{86}Sr = 0.7022 ~ 0.7060 and ^{143}Nd/^{144}Nd = 0.5135 ~ 0.5118; Fig. 5), that could be explained by the peridotite-melt reaction (Tang et al., 2008). On the one hand, the difference between major-element compositions of basaltic melt derived from partial melting of asthenosphere (Fo in olivine ~89) and those of mantle peridotites (Fo in olivine ~92) is relatively small and thus the decrease of olivine Fo in mantle peridotites, caused by the asthenospheric melt-peridotite reaction, is small. On the other hand, the asthenospheric melt-peridotite interaction causes the depletion in Sr-Nd isotopic compositions of mantle peridotites due to the depleted Sr-

Nd isotopic ratios in asthenospheric melts. Possibly, the peridotite-melt interaction could not cause a large variation in Re-Os isotopic system of mantle peridotites because Os isotope systematics for cratonic peridotites appear to be dominantly influenced by the ancient differentiation events that caused them to separate from the convecting mantle, whereas Sr-Nd isotope systematics record later events (Pearson, 1999). Thus, the debate between Os isochron ages (~1.9 Ga) and Sr-Nd isotopic compositions (depleted) of Hannuoba mantle xenoliths can be explained with the fairly recent effect of the peridotite-melt reaction. The abundance of garnet-bearing pyroxenites in Hannuoba xenoliths indicates the presence of peridotite-melt reaction (Liu et al., 2005; Zhang et al., 2009).

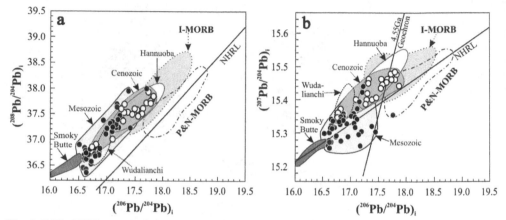

Fig. 6. $^{206}Pb/^{204}Pb_i$ vs. $^{208}Pb/^{204}Pb_i$ and $^{207}Pb/^{204}Pb_i$ diagrams for the basaltic rocks. Data sources: Fields of I-MORB (Indian MORB), P&N-MORB (Pacific & North Atlantic MORB) and NHRL (north hemisphere reference line) (Barry & Kent, 1998; Hart, 1984; Zou et al., 2000), field for Smoky Butte lamproites (Fraser et al., 1985), Wudalianchi basalts (Liu et al., 1994), Hannuoba basalts as in Fig. 5, other data sources and symbols as in Fig. 2.

Similarly, some peridotite xenoliths entrained in the Hannuoba and Fanshi basalts have pyroxenite veins, indicating the presence of peridotite-melt reaction in the mantle lithosphere beneath the Central Zone of the NCC. The variations in isotopic ratios of these xenoliths might indicate the heterogeneity of peridotite-melt reaction (Tang et al., 2011). As a result, the enriched isotopic composition of cpx from the Fanshi and Yangyuan peridotite xenoliths could represent the signatures of old lithospheric mantle, which have experienced/or not such a peridotite-melt reaction.

The existence of old lithospheric mantle beneath the Central Zone during the Cenozoic is also proved by the discovery of mantle olivine xenocrysts in the Xiyang-Pingding basalts (Tang et al., 2004) and high Mg# (Fo≥92) peridotite xenoliths hosted by the Hebi basalts (Zheng et al., 2001), which are interpreted as the relics of old lithospheric mantle. The involvement of old lithospheric mantle in asthenospheric mantle source might well account for the isotopic features of the Cenozoic basalts (Fig. 5). In terms of Sr and Nd elemental contents and isotopic ratios of $^{87}Sr/^{86}Sr_i$ and $^{143}Nd/^{144}Nd_i$, the hypothetical mixing modeling between depleted mantle (DM; Zindler & Hart, 1986; Flower et al., 1998) and old lithospheric mantle (represented by the mantle-derived xenoliths with radiogenic isotopic compositions) reveals that the addition of 4~20% old lithospheric component into the DM will generate the observed Sr-Nd isotopic compositions for these Cenozoic basalts (Fig. 5).

Secular Evolution of Lithospheric Mantle Beneath the Central North China Craton: Implication
from Basaltic Rocks and Their Xenoliths

9

According to the modelling results from the classic, non-modal batch melting equations of Shaw (1970), small degrees of partial melting of a garnet-bearing lherzolitic mantle source are required to explain the REE patterns observed in these basalts (Fig. 7, Tang et al., 2006), which is consistent with the low HREE contents of these Cenozoic basalts. The systematic presence of garnet as a residual phase requires melting depth in excess of 70-80 km, where garnet becomes stable. The results (Fig. 7) also suggest a deeper origin for the Zuoquan and Xiyang-Pingding basalts due to the higher garnet contents in their mantle source than those for the Fanshi-Yingxian basalts, as garnet becomes more with increasing depth.

A lithospheric profile model (Fig. 8c) illustrates the lithospheric evolution and the Cenozoic magmatism in the Central Zone. The Cenozoic tensional regime likely related to the Indian-Eurasian collision (Ren et al., 2002; Liu et al., 2004; Xu et al., 2004) might reactivate old faults, then the old lithospheric mantle was heated by progressively thermo-mechanical erosion processes with the upwelling of asthenosphere. As a result, the base lithosphere was gradually removed by the convecting mantle, forming a mixture of material from the old lithospheric mantle with the magmas from the asthenosphere, which finally produced the Cenozoic basalts through partial melting.

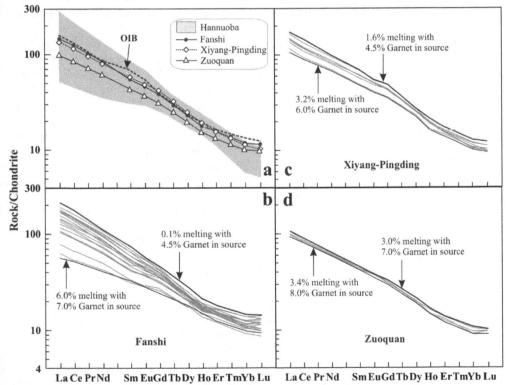

Fig. 7. Chondrite-normalized REE patterns for the Cenozoic basalts (Tang et al., 2006). Mean values of the REE for the basalts (a). Non-modal batch melting models used to approach partial melts for Fanshi (b), Xiyang-Pingding (c) and Zuoquan basalts (d). Data sources: Chondrite (Anders & Grevesse, 1989), OIB (Sun & McDonough, 1989).

5.2 Nature of the Mesozoic lithospheric mantle

Compared with the Cenozoic basalts, the Mesozoic basaltic rocks have obviously higher SiO_2 content with lower FeO^T and TiO_2, and are depleted in HFSE, displaying typical EM1 character in isotopic compositions, which show the clear distinction between their mantle sources.

Element ratios, such as Nb/U, Ce/Nb, Zr/Nb, Ce/Ba and Ce/Pb, are demonstrated to be effective indicators for discriminating mantle source of asthenospheric or lithospheric origin and whether there were subducted materials involved in magma geneses (Salters & Shimizu, 1988; Kelemen et al., 1990; Hofmann, 1997; Turner & Foden, 2001). Plots of trace-element ratios (Fig. 4) show the remarkable differences between Mesozoic and Cenozoic basaltic rocks. Strong depletion in HFSE reveals some similarities of mantle sources between the Mesozoic rocks and arc magma in mantle wedges (Kelemen et al., 1990; Turner & Foden, 2001). Higher Ce/Nb, Zr/Nb, Ba/Nb, but lower Nb/U ratios (Fig. 4) in Mesozoic rocks relative to the Cenozoic basalts indicate that the source for these intrusive rocks are enriched in LREE and Zr relative to the Nb, and depleted in Nb. Their isotopic differences between Mesozoic and Cenozoic basaltic rocks are also obvious (Figs. 5 & 6). These geochemical signatures suggest that the Mesozoic rocks originated from a modified lithospheric mantle, and their low Nb/U ratios (Fig. 4d) and depletion in HFSE (Fig. 3) indicate the involvement of subducted crustal materials in magma geneses (Hofmann, 1997).

Geochemical compositions of the Mesozoic basaltic rocks from the Central Zone indicate that the secular evolution of old cratonic lithospheric mantle underwent processes of modification, which are believed to have originated from the influx of materials with old provenance age, which over time would develop isotopic enrichment (Zhang & Sun, 2002). The Sr-Nd isotopic compositions for these Mesozoic rocks indicate that the source was depleted in Rb but enriched in LREE. Their low Pb isotopic ratios (Fig. 6) define a trend towards the field for Smoke Butte lamproites, which originated from an EMI-like lithospheric mantle. These features, coupled with the clear depletion in HFSE and enrichment in LILE, suggest the involvement of an old component with low Sm/Nd, Rb/Sr and U/Pb ratios. It's the secular evolution of modified lithospheric mantle by old component leads to the striking features of very low ratios of $^{143}Nd/^{144}Nd_i$ (<0.5120) and $^{206}Pb/^{204}Pb_i$ (16.5~17.5), slightly low $^{87}Sr/^{86}Sr_i$ ratios (most = 0.7050~0.7065) of the Mesozoic basaltic rocks from the Central Zone (Figs. 5 & 6).

Mantle xenoliths, discovered in Palaeozoic kimberlites from the NCC, have very restricted Nd isotopic compositions (Fig. 5). In contrast, Nd isotopic compositions for Mesozoic Jinan gabbros, in the centre of the NCC, are slightly lower than those of Palaeozoic kimberlite-borne mantle xenoliths. The interpretation is that their mantle source inherited the characteristics of old lithospheric mantle with slight modification because the significant crustal contamination or AFC process during magma evolution has been excluded (Guo et al., 2001; Zhang et al., 2004a), as shown by their high MgO contents and the lack of a positive correlation of $^{87}Sr/^{86}Sr_i$ with SiO_2 or Mg# in these gabbroic rocks. Similarly, Mesozoic rocks from the Central Zone are lower in Nd isotopic ratios than the Jinan gabbros, indicating that the Mesozoic lithospheric mantle beneath the Central Zone was modified considerably by some mantle enrichment processes. It is interesting to note that the Nd isotopic ratios of the Mesozoic rocks are nearly equal to those of the Mesozoic Zouping gabbros from the centre of the NCC (Fig. 5), and the genesis of the latter are linked to carbonatitic metasomatism of lithospheric mantle (Ying et al., 2005).

Secular Evolution of Lithospheric Mantle Beneath the Central North China Craton: Implication
from Basaltic Rocks and Their Xenoliths

11

On the basis of the above discussions, we propose that carbonatitic and silicic metasomatism may be a suitable candidate for the modification of the old lithospheric mantle beneath the Central Zone. The metasomatised agents should be enriched in LILE and Sr-Nd isotopic, depleted in HFSE and Pb isotopic ratios, and low in Sm/Nd, Rb/Sr and U/Pb ratios, whose geochemical features suggest that they can only be derived from old subducted crustal materials. As yet, there is no clear evidence to explain the occurrence of Phanerozoic subduction/collision in the interior of the NCC, except the Paleoproterozoic collision (~1.8 Ga) between the Eastern Block and the Western Block of the NCC (Gilder et al., 1991; Zhao et al., 2001; Wang et al., 2004). Thus, the carbonatitic and silicic metasomatism for the old lithospheric mantle beneath the Central Zone were probably related to the Paleoproterozoic collision between the two blocks.

5.3 Tectonic and magmatic model

The North China Craton is bounded on the south by the Paleozoic to Triassic Qinling-Dabie-Sulu orogenic belt (Li et al., 1993) and on the north by the Central Asian Orogenic Belt (Şengör et al., 1999; Jahn et al., 2000). The Triassic ages for the Dabie-Sulu UHP rocks in the southern margin of the NCC have been summarized (Zheng et al., 2003). The Central Asian Orogenic Belt formed through a complicated subduction and accretion processes and post-collisional magamtism over a long period of time ranging from the Early Paleozoic through the Triassic (Jahn et al., 2000). These subduction and the subsequent collisions may have affected the stability of the lithospheric mantle beneath the NCC (Zhang et al., 2003 and references therein). The westward subduction of the Pacific plate beneath the Euroasian continent provides the geodynamic setting of back-arc extension for the massive occurrence of Early Cretaceous igneous rocks in the east China continent (Wu et al., 2005). However, these magmatism just took place in Early Cretaceous rather than continuously from Jurassic to present, which requires a thermal pulse to cause the short-lived but large-scale anatexis of thickened lithosphere as a remote response to the Pacific superplume event (Zhao et al., 2005). This event may essentially act as mantle superwelling beneath the Euroasian continent that supply the excess heat to fuse the lithospheric mantle and overlying crust because material contribution of mantle plume hasn't been identified in the contemporaneous igneous rocks from the eastern edge of China continent.

On the basis of the above discussion and previous documents (Zhao et al., 2001, 2010; Zhang and Sun, 2002; Zhang et al., 2003; Wang et al., 2004; Faure et al., 2007; Zheng et al., 2009, 2010), we summarize a tectonic and magmatic model for the secular evolution of the lithospheric mantle beneath the Taihang Mountains (Fig. 8):

1. In the Late Archean to Paleoproterozoic, the Western Block (Zhao et al., 2001, 2010; Wang et al., 2004) and/or Eastern Block (Faure et al., 2007; Zheng et al., 2009) was subducted beneath the Central Zone with subduction of old continental and oceanic crustal component to mantle depths. Meanwhile, sedimentary rocks of the Eastern and Western Blocks were thrust over the Central Zone, which caused crustal-scale folding, thrusting and metamorphism, associated with the initial metasomatism of old lithospheric mantle by carbonatitic and silicic agents. At ~1.85 Ga, the orogenic belt suffered post-collision extensional collapse, which was associated with the subducted slab detachment and the development of the mantle metasomatism for the old lithospheric mantle. As a result, the Paleoproterozoic collision between the Eastern and Western Blocks led to the assembly of the NCC and the modification of old lithospheric

mantle by carbonatitic and silicic metasomatism (Fig. 8a). According to recent studies (Zhao et al., 2010; Zheng et al., 2010), the direction of subduction polarity in the Central Zone has still not been resolved. Whether the subduction polarity is westward or eastward the event(s) had led to the modification of the old lithospheric mantle by subducted crustal materials.

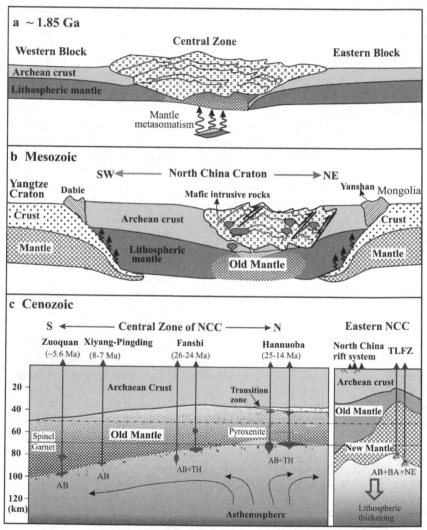

Fig. 8. Schematic cartoons of tectonic and magmatic model, showing the secular evolution of lithospheric mantle beneath the Central Zone of the NCC (a~c). Sketch map (a) is modified from Zhao et al. (2001), Wang et al. (2004) and Zheng et al. (2009); map (b) is modified from Zhang et al. (2003); map (c) is modified from Tang et al. (2006) and Menzies and Xu (1998). AB, alkaline basalt; AOB, alkaline olivine basalt; BA, Basanite; NE, nephelinite; OTH, olivine tholeiite. See text for the detail.

Secular Evolution of Lithospheric Mantle Beneath the Central North China Craton: Implication
from Basaltic Rocks and Their Xenoliths

13

2. Subduction and collisions along the northern and southern margins of the North China Craton especially in Triassic initiated the cracking in the NCC interior. Late Mesozoic lithospheric thinning and mafic magmatism might have occurred with the upwelling of the asthenosphere probably also as a remote response to the Pacific superplume event (Zhao et al., 2005). With the change from convergent to extensional regime, the Mesozoic intrusive rocks might be generated by the partial melting of the metasomatised old lithospheric mantle beneath the Taihang Mountains (Fig. 8b).

3. With the continental extension in the Central Zone, possibly related to the Early Tertiary Indian-Eurasian collision, the Cenozoic basalts were produced by the decompression melting of asthenosphere and the interaction between asthenospheric magmas and old lithospheric mantle (Tang et al., 2006). The substantive existence of old lithospheric mantle with some modification by asthenospheric melt in the Central Zone is remarkably different from the Cenozoic lithospheric accretion in the eastern North China Craton (Fig. 8c).

6. Conclusion

Geochemical compositions indicate that the Mesozoic basaltic rocks from the Central Zone originated from lithospheric mantle, which was enriched in LREE, LILE and Sr-Nd isotopic ratios and depleted in HFSE and Pb isotopic compositions. The lithospheric mantle with these geochemical features had been probably produced by the modification of old cratonic lithospheric mantle with carbonatitic and silicic metasomatism, which were mainly derived from the subducted crustal materials during the Paleoproterozoic collision between the Eastern and Western blocks of the NCC.

Cenozoic basalts from the Central Zone were generated from the partial melting of asthenospheric mantle with/without some contributions of old lithospheric mantle during continental extension, which might be related to the Early Tertiary Indian-Eurasian collision. In conjunction with the data of mantle peridotite xenoliths, the Cenozoic lithospheric mantle has inherited the isotopic features of old lithosphere mantle in spite of some signatures of the modification by the asthenospheric melt-peridotite reaction.

7. Acknowledgement

We are grateful to two anonymous referees and the book editor for their constructive comments that significantly improved the manuscript. This work was financially supported by the Natural Science Foundation of China (91014007, 41073028 and 40773026).

8. References

Arndt, N. T. & Christensen, U. (1992) The role of lithospheric mantle in continental flood volcanism: Thermal and geochemical constraints. *J. Geophy. Res.* 97, 10967-10981.

Anders, E. & Grevesse, N. (1989) Abundances of the elements: meteoritic and solar. *Geochim. Cosmochim. Acta* 53, 197-214.

Barry, T. L. & Kent, R. W. (1998) Cenozoic magmatism in Mongolia and the origin of central and east Asian basalts. *Mantle Dynamics and Plate Interactions in East Asia, Geodynamics Series 27,* (Flower, M. F. J., Chung, S. L., Lo, C. H. and Lee, T. Y., eds.) 347-364, AGU, Washington.

Basu, A. R., Wang, J. W., Huang, W. K., Xie, G. H. & Tatsumoto, M. (1991) Major element, REE, and Pb, Nd, and Sr isotopic geochemistry of Cenozoic volcanic rocks of eastern China: implications for their origin from suboceanic-type mantle reservoirs. *Earth Planet. Sci. Lett.* 105, 149-169.

Cai, J. H., Yan, G. H., Chang, Z. S., Wang X. F., Shao, H. X. & Chu, Z. Y. (2003) Petrological and geochemical characteristics of the Wanganzhen complex and discussion on its genesis. *Acta Petrologica Sinica* 19(1), 81-92 (in Chinese with English abstract).

Chen, B., Jahn, B. M. & Zhai, M. G. (2004) Petrogenesis of the Mesozoic intrusive complexes from the southern Taihang Orogen, North China Craton: element and Sr-Nd-Pb isotopic constraints. *Contrib. Mineral. Petrol.* 148, 489-501.

Chen, B. & Zhai, M. G. (2003) Geochemistry of late Mesozoic lamprophyre dykes from the Taihang Mountains, north China, and implications for the sub-continental lithospheric mantle. *Geol. Mag.* 140(1), 87-93.

Chen, B., Zhai, M. G. & Shao, J. A. (2003) Petrogenesis and significance of the Mesozoic North Taihang complex: major and trace element evidence. *Science in China (series D)* 46(9), 48-60.

Deng, J. F., Mo, X. X., Zhao, H. L., Wu, Z. X., Luo, Z. H. & Su, S. G. (2004) A new model for the dynamic evolution of Chinese lithosphere: 'continental roots-plume tectonics'. *Earth Sci. Rev.* 65, 223-275.

Depaolo, D. J. & Daley, E. E. (2000) Neodymium isotopes in basalts of the southwest basin and range and the lithospheric thinning during continental extension. *Chem. Geol.* 169, 157-185.

Dong, J.H., Chen, B., & Zhou, L. (2003) The genesis of fushan terrane in the southern Taihang mountains. *Progress in Nature Sci.* 13, 767-74. (in Chinese)

Falloon, T.J., Green, D.H., Harton, C.J., & Harris, K.J. (1988) Anhydrous partial melting of a fertile and depleted peridotite from 2 to 30 kb and application to basalt petrogenesis. *J. Petrol.* 29, 1257-1282.

Fan, W. M., Zhang, H. F., Baker, J., Jarvis, K. E., Mason, P. R. D. & Menzies, M. A. (2000) On and off the north China craton: Where is the Archaean keel? *J. Petrol.* 41, 933-950.

Faure, M., Trap, P., Lin, W., Monié, P. & Bruguier, O. (2007) Polyorogenic evolution of the paleoproterozoic trans-north china belt. *Episodes* 30, 96-107.

Flower, M., Tamaki, K. & Hoang, N. (1998) *Mantle Dynamics and Plate Interactions in East Asia, Geodynamics Series 27*, (Flower, M. F. J., Chung, S. L., Lo, C. H. and Lee, T. Y., eds.) 67-88, AGU, Washington.

Fraser, K. J., Hawkesworth, C. J., Erland, A. J., Mitchell, R. H. & Scott-Smith, B. H. (1985) Sr, Nd and Pb isotope and minor element geochemistry of lamproites and kimberlites. *Earth Planet. Sci. Lett.* 76, 57-70.

Gao, S., Rudnick, R. L., Carlson, R. W., Mcdonough, W. F. & Liu, Y. S. (2002) Re-Os evidence for replacement of ancient mantle lithosphere beneath the North China craton. *Earth Planet. Sci. Lett.* 198, 307-322.

Gao, S., Rudnick, R. L., Yuan, H. L., Liu, X. M., Liu, Y. S., Xu, W. L., Ling, W. L., Ayers, J., Wang, X. C. & Wang, Q. H. (2004) Recycling lower continental crust in the North China craton. *Nature* 432, 892-897.

Gao, S., Zhang, B. R., Jin, Z. M., Kern, H., Luo, T. C. & Zhao, Z. D. (1998) How mafic is the lower continental crust? *Earth Planet. Sci. Lett.* 161, 101-117.

Secular Evolution of Lithospheric Mantle Beneath the Central North China Craton: Implication
from Basaltic Rocks and Their Xenoliths

15

Gilder, S. A., Keller, G. R., Luo, M. & Goodell, P. C. (1991) Eastern Asia and the western Pacific: Timing and spatial distribution of rifting in China. *Tectonophysics* 197, 225-243.

Griffin, W. L., O'Reilly, S. Y. & Ryan, C. G. (1992) Composition and thermal structure of the lithosphere beneath South Africa, Siberia and China: proton microprobe studies. *Proceedings, International Symposium on Cenozoic Volcanic Rocks and Deep-seated Xenoliths of China and its Environs*, Beijing, August, 1-20.

Griffin, W. L., Zhang, A. D., O'Reilly, S. Y. & Ryan, C. G. (1998) Phanerozoic evolution of the lithosphere beneath the Sino-Korean Craton. *Mantle Dynamics and Plate Interactions in East Asia, Geodynamics Series 27*, (Flower, M. F. J., Chung, S. L., Lo, C. H. and Lee, T. Y., eds.) 107-126, AGU, Washington.

Guo, F., Fan, W. M., Wang, Y. J. & Lin, G. (2001) Late Mesozoic mafic intrusive complexes in North China Block: constraints on the nature of subcontinental lithospheric mantle. *Phys. Chem. Earth (A) 26*, 759-771.

Guo, F., Fan, W. M., Wang, Y. J. & Lin, G. (2003) Geochemistry of late Mesozoic mafic magmatism in west Shandong Province, eastern China: Characterizing the lost lithospheric mantle beneath the North China Block. *Geochem. J. 37*, 63-77.

Hart, S. R. (1984) A large scale isotope anomaly in the southern hemisphere mantle. *Nature* 309, 753-757.

Herzberg, C. (2006) Petrology and thermal structure of the Hawaiian plume from Mauna Kea volcano. *Nature 444*, 605-609.

Hofmann, A. W. (1997) Early evolution of continents. *Science* 275, 498-499.

Hofmann, A.W. (2003) Sampling mantle heterogeneity through oceanic basalts: isotopes and trace elements. *The Mantle and Core* (Carlson, R. W. eds.) Vol.2 *Treatise on Geochemistry* (Holland, H. D. and Turekian, K. K. eds.), Elsevier-Pergamon, Oxford. pp. 61-101.

Irving, T. N. & Baragar, W. R. A. (1971) A guide to the chemical classification of common volcanic rocks. *Can. J. Earth Sci 8*, 523-548.

Jahn, B. M. & Zhang, Z. Q. (1984) Archean granulite gneisses from eastern Hebei province, China: rare earth geochemistry and tectonic implications. *Contrib. Mineral. Petrol.* 85, 224-243.

Jahn, B. M., Auvray, B., Shen, Q. H., Liu, D. Y., Zhang, Z. Q., Dong, Y. J., Ye, X. J., Zhang, Q. Z., Cornichet, J. & Mace, J. (1988) Archean crustal evolution in China: the Taishan complex, and evidence for Juvenile crustal addition from long-term depleted mantle. *Precam. Res.* 38, 381-403.

Jahn, B.M., Wu, F.Y. & Chen, B. (2000) Granitoids of the Central Asian Orogenic Belt and Continental Growth in the Phanerozoic. *Trans. R. Soc. Edinburgh Earth Sci.* 91, 181-193.

Kelemen, P. D., Johnson, K. T. M., Kinzler, R. J. & Irving, A. J. (1990) High-field-strength element depletions in arc basalts due to mantle-magma interaction. *Nature 345*, 521-524.

Le Bas, M., Le Maitre, R. W., Strekeisen, A. & Zanettin, B. (1986) A chemical classification of volcanic rocks based on the total alkali-silica diagram. *J. Petrol.* 27, 745-750.

Li, S. G., Xiao, Y. L., Liou, D. L., Chen, Y. Z., Ge, N. J., Zhang, Z. Q., Sun, S. S., Cong, B. L., Zhang, R. Y., Hart, S. R. & Wang, S. S. (1993) Collision of the North China and

Yangtze Blocks and formation of coesite-bearing eclogites-Timing and processes. *Chem. Geol.* 109, 89-111.

Liu, C. Q., Masuda, A. & Xie, G. H. (1994) Major- and Trace-element compositions of Cenozoic basalts in eastern China: Petrogenesis and mantle source. *Chem. Geol.* 114, 19-42.

Liu, D. Y., Nutman, A. P., Compston, W., Wu, J. S. & Shen, Q. H. (1992a) Remnants of 3800 Ma crust in the Chinese part of the Sino-Korean craton. *Geology* 20, 339-342.

Liu, M., Cui, X. & Liu, F. (2004) Cenozoic rifting and volcanism in eastern China: a mantle dynamic link to the Indo-Asian collision? *Tectonophysics* 393, 29-42.

Liu, R. X., Chen, W. J., Sun, J. Z. & Li, D. M. (1992b) The K-Ar age and tectonic environment of Cenozoic volcanic rock in China. *The age and geochemistry of Cenozoic volcanic rock in China* (Liu, R. X. eds.) 1-43, Seismologic Press. (in Chinese).

Liu, Y. S., Gao, S., Lee, C.-T. A., Hu, S. H., Liu, X. M. & Yuan, H. L. (2005) Melt-peridotite interactions: Links between garnet pyroxenite and high-Mg# signature of continental crust. *Earth Planet. Sci. Lett.* 234, 39-57.

Lu, F. X., Zhao, L., Deng, J. F. & Zheng, J. P. (1995) The discussion on the ages of kimberlitic magmatic activity in North China platform. *Acta Geologica Sinica* 11, 365-374 (in Chinese with English abstract).

Ma, J. L., & Xu, Y. G. (2006) Old EM1-type enriched mantle under the middle North China Craton as indicated by Sr and Nd isotopes of mantle xenoliths from Yangyuan, Hebei Province. *Chinese Science Bulletin* 51, 1343-1349.

Ma, X. (1989) Atlas of active faults in China. Seismologic Press, 120 pp. Beijing.

McDonough, W. F. & Sun, S. S. (1995) The composition of the earth. *Chem. Geol.* 120, 223-253.

Menzies, M. A., Fan, W. M. & Zhang, M. (1993) Palaeozoic and Cenozoic lithoprobes and the loss of > 120 km of Archaean lithosphere, Sino-Korean craton, China. *Magmatic processes and plate tectonics* (Prichard, H. M., Alabaster, T., Harris, N. B. W. & Neary, C. R., eds.), *Geol. Soc. Spec. Publ.* 76, 71-81.

Menzies, M. A. & Xu, Y. G. (1998) Geodynamics of the North China Craton. *Mantle Dynamics and Plate Interactions in East Asia, Geodynamics Series* 27, (Flower, M. F. J., Chung, S. L., Lo, C. H. & Lee, T. Y., eds.) 155-165, AGU, Washington.

Pearson, D. G. (1999) Evolution of cratonic lithospheric mantle: an isotopic perspective. *Mantle Petrology: Field Observations and High-Pressure Experimentation: A Tribute to Francis R. (Joe) Boyd* (Fei, Y., Bertka, C. M. & Mysen, B. O. eds.). Geochem. Soc. Spec. Publ. 6, 57-78.

Peng, T. P., Wang, Y. J., Fan, W. M., Peng, B. X. & Guo, F. (2004) SHRIMP zircon U-Pb geochronology of the diorites for the southern Taihang Mountains in Central North Interior and its petrogenesis. *Acta Petrologica Sinica* 20(5), 1253-1262 (in Chinese with English abstract).

Peng, Z. C., Zartman, R. E., Futa, K. & Chen, D. G. (1986) Pb-, Sr- and Nd- isotopic systematics and chemical characteristic of Cenozoic basalts, eastern China. *Chem. Geol.* 59, 3-33.

Ren, J., Tamaki, K., Li, S. & Zhang, J. (2002) Late Mesozoic and Cenozoic rifting and its dynamic setting in eastern China and adjacent areas. *Tectonophysics* 344, 175-205.

Rudnick, R. L. & Gao, S. (2003) Composition of the continental crust. *The Crust* (Rudnick, R. L. eds.), *Treatise on Geochemistry* (Holland, H. D. & Turekian, K. K. eds.), Vol. 3, 1-64. Elsevier-Pergamon, Oxford.

Secular Evolution of Lithospheric Mantle Beneath the Central North China Craton: Implication
from Basaltic Rocks and Their Xenoliths

17

Rudnick, R. L., Gao, S., Ling, W. L., Liu, Y. S. & McDonough, W. F. (2004) Petrology and geochemistry of spinel peridotite xenoliths from Hannuoba and Qixia, North China Craton. *Lithos* 77, 609-637.

Salters, V. J. M. & Shimizu, N. (1988) World-wide occurrence of HFSE-depleted mantle. *Geochim. Cosmochim. Acta* 52, 2177-2182.

Şengör, A. M. C., Natal'in, B. A. & Burtman, V. S. (1999) Evolution of the altaid tectonic collage and Palaeozoic crustal growth in Eurasia. *Nature* 364, 299-307.

Shaw, D.M. (1970) Trace elements fractionation during anatexis. *Geochim. Cosmochim. Acta* 34, 237-243.

Song, Y., Frey, F. A. & Zhi, X. (1990) Isotopic characteristics of Hannuoba basalts, eastern China: Implications for their petrogenesis and the composition of subcontinental mantle. *Chem. Geol.* 88, 35-52.

Sun, S. S. & McDonough, W. F. (1989) Chemical and isotopic systematic of oceanic basalt: implication for mantle composition and processes. *Magmatism in the oceanic basins* (Saunders, A. D. & Norry, M. J., eds.), Spec. Publ. Geol. Soc. London. 42, 313-346.

Tan, D.J., & Lin, J.Q. (1994) Mesozoic potassium magma province on north china platform.: The seismological press, 184pp. Beijing (in Chinese).

Tang, Y. J., Zhang, H. F. & Ying, J. F. (2004) High-Mg olivine xenocrysts entrained in Cenozoic basalts in central Taihang Mountains: relics of old lithospheric mantle. *Acta Petrologica Sinica* 20(5), 1243-1252 (in Chinese with English abstract).

Tang, Y. J., Zhang, H. F. & Ying, J. F. (2006) Asthenosphere-lithospheric mantle interaction in an extensional regime: implication from the geochemistry of Cenozoic basalts from Taihang Mountains, North China Craton. *Chem.Geol.* 233, 309-327.

Tang, Y. J., Zhang, H. F., Nakamura, E., Moriguti, T., Kobayashi, K. & Ying, J. F. (2007) Lithium isotopic systematics of peridotite xenoliths from Hannuoba, North China Craton: implications for melt-rock interaction in the considerably thinned lithospheric mantle. *Geochimica et Cosmochimica Acta*, 71, 4327-4341.

Tang, Y. J., Zhang, H. F., Ying, J. F., Zhang, J. & Liu, X. M. (2008) Refertilization of ancient lithospheric mantle beneath the central North China Craton: Evidence from petrology and geochemistry of peridotite xenoliths. *Lithos*, 101, 435-452.

Tang, Y. J., Zhang, H. F., Nakamura, E. & Ying, J. F. (2011) Multistage melt/fluid-peridotite interactions in the refertilized lithospheric mantle beneath the North China Craton: Constraints from the Li-Sr-Nd isotopic disequilibrium between minerals of peridotite xenoliths. *Contributions to Mineralogy and Petrology*, doi: 10.1007/s00410-010-0568-1.

Tatsumoto, M., Basu, A.R., Huang, W. K., Wang, J. W., & Xie, G. H. (1992) Sr, Nd, and Pb isotopes of ultramafic xenoliths in volcanic-rocks of eastern China: enriched components EMI and EMII in subcontinental lithosphere. *Earth and Planetary Science Letters* 113, 107-128.

Tu, K., Flower, M. F., Carlson, R. W., Zhang, M. & Xie, G. H. (1991) Sr, Nd, and Pb isotopic compositions of Hainan basalts (south China): implications for a subcontinental lithosphere Dupal source. *Geology* 19, 567-569.

Turner, S. & Foden, J. (2001) U, Th and Ra disequilibria, Sr, Nd and Pb isotope and trace element variations in Sunda arc lavas: predominance of a subducted sediment component. *Contrib. Mineral. Petrol.* 142, 43-57.

Turner, S. & Hawkesworth, C. (1995) The nature of the sub-continental mantle: constraints from the major element composition of continental flood basalts. *Chem. Geol.* 120, 295-314.

Wang, Y. J., Fan, W. M. & Zhang, Y. H. (2004) Geochemical, ^{40}Ar/^{39}Ar geochronological and Sr-Nd isotopic constraints on the origin of Paleoproterozoic mafic dikes from the southern Taihang Mountains and implications for the ca. 1800 Ma event of the North China Craton. *Precam. Res.* 135(1-2), 55-79.

Wilde, S. A., Zhou, X. H., Nemchin, A. A. & Sun, M. (2003) Mesozoic crust-mantle beneath the North China craton: a consequence of the dispersal of Gondwanaland and accretion of Asia. *Geology* 31, 817-820.

Wu, F. Y., Lin, J. Q., Wilde, S. A., Zhang, X. O. & Yang, J. H. (2005) Nature and significance of the Early Cretaceous giant igneous event in eastern China. *Earth Planet. Sci. Lett.* 233, 103-119.

Wu, F. Y., Walker, R. J., Ren, X. W., Sun, D. Y. & Zhou, X. H. (2003) Osmium isotopic constraints on the age of lithospheric mantle beneath northeastern China. *Chem. Geol.* 196, 107-129.

Xie, G. H. & Wang, J. W. (1992) The geochemistry of Hannuoba basalts and their ultra-mafic xenoliths. *The age and geochemistry of Cenozoic volcanic rock in China* (Liu, R. X., eds.), 149-170, Seismologic Press. (in Chinese).

Xu, W.L. and Lin, J.Q. (1991) The discovery and study of mantle-derived dunite inclusions in hornblende-diorite in the Handan-Xingtai area, Hebei. *Acta Geologica Sinica*, 65, 33-41 (in Chinese with English abstract).

Xu, Y. G. (2001) Thermo-tectonic destruction of the Archean lithospheric keel beneath the Sino-Korean Craton in China: Evidence, timing and mechanism. *Phys Chem Earth (A)* 26, 747-757.

Xu, Y. G., Chung, S. L., Ma, J. L. & Shi, L. B. (2004) Contrasting Cenozoic lithospheric evolution and architecture in the western and eastern Sino-Korean craton: constrains from geochemistry of basalts and mantle xenoliths. *The Journal of Geology* 112, 593-605.

Ye, H., Zhang, B. T. & Ma, F. (1987) The Cenozoic tectonic evolution of the great North China: two types of rifting and crustal necking in the great North China and their tectonic implications. *Tectonophysics* 133, 217-227.

Ying, J. F., Zhou, X. H. & Zhang, H. F. (2004) Geochemical and isotopic investigation of the Laiwu-Zibo carbonatites from western Shandong Province, China and implications for their petrogenesis and enriched mantle source. *Lithos* 75, 413-426.

Ying, J. F., Zhou, X. H. & Zhang, H. F. (2005) The Geochemical variations of mid-Cretaceous lavas across western Shandong Province, China and their tectonic implications. *Intl. J. Earth Sci.* (in press).

Zhang, H.F. (2005) Transformation of lithospheric mantle through peridotite-melt reaction: A case of Sino-Korean Craton. *Earth Planet. Sci. Lett.* 237, 768-780.

Zhang, H. F. & Sun, M. (2002) Geochemistry of Mesozoic basalts and mafic dikes, southeastern north China craton, and tectonic implications. *Intl. Geol. Rev.* 44, 370-382.

Zhang, H. F., Sun, M., Zhou, X. H., Fan, W. M., Zai, M. G. & Ying, J. F. (2002) Mesozoic lithosphere destruction beneath the North China Craton: evidence from major-,

Secular Evolution of Lithospheric Mantle Beneath the Central North China Craton: Implication
from Basaltic Rocks and Their Xenoliths

19

trace-element and Sr-Nd-Pb isotope studies of Fangcheng basalts. *Contrib. Mineral. Petrol.* 144, 241-253.

Zhang, H. F., Sun, M., Zhou, X. H., Zhou, M. F., Fan, W. M. & Zheng, J. P. (2003) Secular evolution of the lithosphere beneath the eastern North China Craton: Evidence from Mesozoic basalts and high-Mg andesites. *Geochim. Cosmochim. Acta* 67, 4373-4387.

Zhang, H. F., Sun, M., Zhou, M. F., Fan, W. M., Zhou, X. H. & Zhai, M. G. (2004a) Highly heterogeneous late Mesozoic lithospheric mantle beneath the north China Craton: evidence from Sr-Nd-Pb isotopic systematics of mafic igneous rocks. *Geol. Mag.* 141(1), 55-62.

Zhang, H. F., Ying, J. F., Xu, P. & Ma, Y. G. (2004b) Mantle olivine xenocrysts entrained in Mesozoic basalts from the North China craton: implication for replacement process of lithospheric mantle. *Chinese Science Bulletin* 49(9), 961-966.

Zhang, H. F., Sun, M., Zhou, X. H. & Ying, J. F. (2005) Geochemical constraints on the origin of Mesozoic alkaline intrusive complexes from the North China Craton and tectonic implications. *Lithos* 81, 297-317.

Zhang, H. F., Goldstein, S., Zhou, X. H., Sun, M., Zheng, J. P. & Cai, Y. (2008) Evolution of subcontinental lithospheric mantle beneath eastern China: Re-Os isotopic evidence from mantle xenoliths in Paleozoic kimberlites and Mesozoic basalts. *Contributions to Mineralogy and Petrology*, 155, 271-293.

Zhang, H. F., Goldstein, S. L., Zhou, X. H., Sun, M. & Cai, Y. (2009) Comprehensive refertilization of lithospheric mantle beneath the North China Craton: further Os-Sr-Nd isotopic constraints. *Journal of the Geological Society, London*, 166, 249-259.

Zhang, M., Zhou, X. H. & Zhang, J. B. (1998) Nature of the lithospheric mantle beneath NE China: Evidence from potassic volcanic rocks and mantle xenoliths. *Mantle Dynamics and Plate Interactions in East Asia, Geodynamics Series 27*, (Flower, M. F. J., Chung, S. L., Lo, C. H. & Lee, T. Y., eds.) 197-219, AGU, Washington.

Zhao, G., Wilde, S. A., Cawood, P. A. & Lu, L. (1999) Tectonothermal history of the basement rocks in the western zone of the North China Craton and its tectonic implications. *Tectonophysics* 310, 37-53.

Zhao, G. C., Wilde, S. A., Cawood, P. A. & Sun, M. (2001) Archean blocks and their boundaries in the North China Craton: lithological, geochemical, structural and P-T path constraints and tectonic evolution. *Precam. Res.* 107, 45-73.

Zhao, G., Wilde, S. A. & Zhang, J. (2010) New evidence from seismic imaging for subduction during assembly of the north china craton: Comment. *Geology*, 38(4), e206.

Zhao, Z. F., Zheng, Y. F., Wei, C. S., Wu, Y. B., Chen, F. K. & Jahn, B. M. (2005) Zircon U-Pb age, element and C-O isotope geochemistry of post-collisional mafic-ultramafic rocks from the Dabie orogen in east-central China. *Lithos* 83, 1-28.

Zheng, J. P., O'Reilly, S. Y., Griffin, W., Lu, F. X., Zhang, M. & Pearson, N. (2001) Relict refractory mantle beneath the eastern North China block: significance for lithosphere evolution. *Lithos* 57, 43-66.

Zheng, T. Y., Zhao L. & Zhu R. X. (2009) New evidence for subduction during assembly of the North China Craton. *Geology* 37: 395-398.

Zheng, T. Y., Zhao, L. & Zhu, R. X. (2010) New evidence from seismic imaging for subduction during assembly of the north china craton: Reply. *Geology*, 38(4), e207.

Zheng, Y. F., Fu, B., Gong, B. & Li, L. (2003) Stable isotope geochemistry of ultrahigh pressure metamorphic rock from the dabie-sulu orogen in china: Implications for geodynamics and fluid regime. *Earth Sci. Rev.* 62, 105-61.

Zhi, X. C., Song, Y., Frey, F. A., Feng, J. L. & Zhai, M. Z. (1990) Geochemistry of Hannuoba basalts, eastern China: constraints on the origin of continental alkalic and tholeiitic basalt. *Chem. Geol.* 88, 1-33.

Zhou, X. H. & Armstrong, R. L. (1982) Cenozoic volcanic rocks of eastern China-secular and geographic trends in chemistry and strontium isotopic composition. *Earth Planet. Sci. Lett.* 58, 301-329.

Zindler, A. & Hart, S. R. (1986) Chemical geodynamics. *Annu. Rev. Earth Planet. Sci.* 14, 493-571.

Zou, H. B., Zindler, A., Xu, X. S. & Qi, Q. (2000) Major, trace element, and Nd, Sr and Pb isotope studies of Cenozoic basalts in SE China: mantle sources, regional variations, and tectonic significance. *Chem. Geol.* 171, 33-47.

Petrologic Study of Explosive Pyroclastic Eruption Stage in Shirataka Volcano, NE Japan: Synchronized Eruption of Multiple Magma Chambers

Masao Ban[1], Shiho Hirotani[2], Osamu Ishizuka[2] and Naoyoshi Iwata[1]
[1]Department of Earth and Environmental Sciences, Yamagata University,
[2]Geological Survey of Japan/AIST
Japan

1. Introduction

Some of detailed petrologic studies on rock samples of middle to large sized explosive pyroclastic eruptions recently revealed that the eruptions were caused by simultaneous eruption of multiple distinct magma chambers beneath the volcanoes (e.g., Nakagawa et al. 2003: Shane et al. 2007). It is very important to examine the genetic relationships among the magmas to understand the magma feeding system which caused such explosive eruptions. The explosive pyroclastic eruption stage in Shirataka volcano, NE Japan (Fig. 1) is one of potential candidates for such kind of researches. The aim of this study is to reveal the magma feeding system beneath Shirataka volcano in the explosive pyroclastic eruption stage and examine the genetic relationships among magmas involved in the explosive eruption.

2. Geologic outline of the Shirataka volcano

The northeast Japan arc (Fig. 1) is one of the representative island arcs. The volume distribution of volcanic products, excluding caldera-related felsic rocks, clearly reveals the existence of two volcanic chains: the frontal row (Nasu volcanic zone) and the back arc row (Chokai volcanic zone) (e.g., Kawano et al. 1961; Tatsumi and Eggins 1995). The Shirataka volcano is situated in the back arc row (Chokai volcanic zone) (Fig. 1). The geologic studies of the Shirataka volcanoes have been performed by many researchers (e.g., Mimura & Kanno, 2000; Ishii & Saito, 1997). Based on the results of these studies, the activity of the Shirataka volcano (0.9-0.7 Ma) is summarized to three stages; strato-cone building, explosive pyroclastic eruption, and lava-dome + block-and-ash flow forming stages (Table 1). The Kokuzo lava, the Numata pumice flow deposit, and Hagino block-and-ash flow deposit + four lava domes (Shiratakayama, Kitsunegoe, Nishikuromoriyama lava dome group + Higashikuromoriyama lava dome) were formed in each stage. Subsequently, the Shirataka volcano collapsed, creating a horseshoe-shaped caldera of ca. 4 km diameter (Yagi et al., 2005).

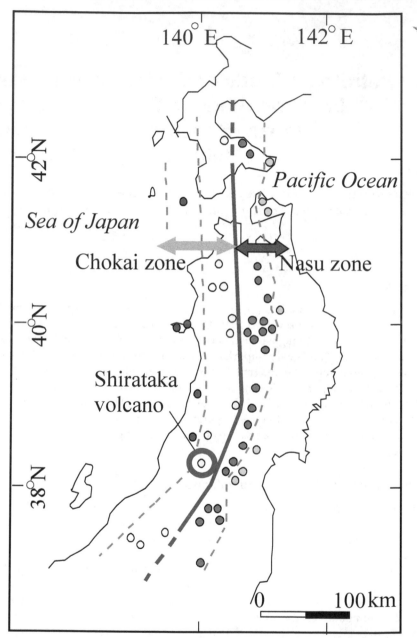

Fig. 1. Location of the Shirataka volcano in the northeast Japan arc. Circles represent Quaternary stratovolcanoes in the northeast Japan arc. The distribution of the volcanoes is from Nakagawa et al. (1988). The distribution of the volcanoes are divided into four volcanic zones: Aoso-Osore (green circle), Sekiryo (orange circle), Moriyoshi (yellow circle), and Chokai (purple circle) zones from trench to rear arc side (Nakagawa et al., 1988).

Petrologic Study of Explosive Pyroclastic Eruption Stage in Shirataka Volcano, NE Japan: Synchronized Eruption
of Multiple Magma Chambers

23

stage 3: lava-dome & block-and-ash flow forming stage

Four lava domes* ⇒ medium-K; low-Cr type

Hagino block-and -ash flow deposit ⇒ medium-K; high- & low-Cr types

stage 2: explosive pyroclastic eruption stage

Numata pumice flow deposit ⇒ medium-K; low-Cr type

stage 1: strato-cone building stage

Kokuzo lava ⇒ medium-K; high- & low Cr types
low-K; high- & low Cr types

Table 1. Simplified stratigraphic relationships and rock series & types of the eruptive products form the Shirataka volcano. *Four lava domes, the Shiratakayama, Kitsunegoe, Nishikuromoriyama lava domes group + Higashikuromoriyama lava dome

3. Petrologic outline of eruptive products from the Shirataka volcano

3.1 Petrography

The eruptive products of the Shirataka volcano are andesitic to dacitic lavas in the first and third stages, but the products are mainly pumice with minor amount of scoria in the second explosive pyroclastic eruption stage. The banded pumice, composed of white pumice and black scoria parts, is rarely observed in the second stage. Mafic magmatic inclusions (~30 cm) are observed in all lavas (Hirotani & Ban, 2006; Hirotani et al., 2009). These are thought to be quenched products of the mafic magma (e.g., Eichelberger, 1980; Bacon, 1986). According to Hirotani & Ban (2006) and Hirotani et al. (2009), the lavas, pumice, and scoria are porphyritic (total phenocryst: ca. 23-40 vol.%) andesitic to dacitic (57.2-65.7 wt% SiO_2), with orthopyroxene, clinopyroxene, plagioclase, and Fe-Ti oxide, with or without hornblende, quartz and olivine phenocrysts. The mafic inclusions are light to dark gray colored, usually rounded in form, and moderately vesiculated, are porphyritic basalt to andesite (48.3-58.5 wt% SiO_2). The total volume of phenocrysts is <ca. 5 vol.% in basalt, and ca. 10 vol.% in basaltic-andesite to andesite, with orthopyroxene, clinopyroxene, and plagioclase, and with or without olivine and quartz phenocrysts. The groundmass has quenched diktytaxitic texture (Bacon, 1986).

3.2 Petrographic evidence for magma mixing and variety of the phenocrysts

The mafic inclusions and disequilibrium phenocryst assemblages (Sakuyama, 1981) such as Mg-rich olivine co-existing with quartz and/or magnesio-hornblende are observed in most of the samples. These features suggest that most of the rocks from Shirataka volcano were formed by magma mixing/mingling (Hirotani & Ban, 2006; Hirotani et al., 2009). Correspondingly, the chemical compositions of phenocrystic minerals show variability. Based on the chemical compositions with textural features, Hirotani & Ban (2006) grouped phenocrystic minerals into several groups. The phenocrysts form different groups cannot co-exist in equilibrium each other by the point of view of their chemical compositions. In the cases of the stages 1 and 3 products, the groups are named to population I, II, and III by Hirotani et al., (2009). Here we refer "population" as "group". Group I includes An-rich plagioclase, Mg-rich olivine and Mg-rich clinopyroxene. Most of these phenocrysts show compositional normal zoning. Group II includes An-poor plagioclase, Mg-poor orthopyroxene, Mg-poor clinopyroxene, hornblende, quartz, titanomagnetite, and ilmenite. Most of the phenocrysts in this population except for Fe-Ti oxides, show compositional reverse zoning and dissolution textures, such as sieved textures (Tsuchiyama, 1985) or embayed form. Group III is made of smaller phenocrysts of plagioclase, olivine, orthopyroxene, and clinopyroxene. These are interpreted to reflect relict cores from a mafic magma, relict cores from a silicic magma, and minerals growing in the hybrid magma caused by mixing the two magmas precipitated the crystal cores of the groups I and II (Hirotani & Ban 2006).

3.3 Rock series and type of eruptive products of the Shirataka volcano

Hirotani & Ban (2006) presented major and trace elements (Ba, Rb, Sr, Zr, Nb, Y, V, Cr, and Ni) compositional data for the studied samples from all stages of the Shirataka volcano. All samples of the Shirataka volcano plot within calc-alkaline series, and most of them belong to medium-K by the definition of Gill (1981), but the lavas and mafic inclusions from the earlier phase of the cone building stage belong to low-K series (Hirotani and Ban, 2006; Hirotani et al., 2009) (Fig. 2a). All of samples from stage 2 belong to medium-K series, but the pumice has slightly lower K_2O contents than the scoria.

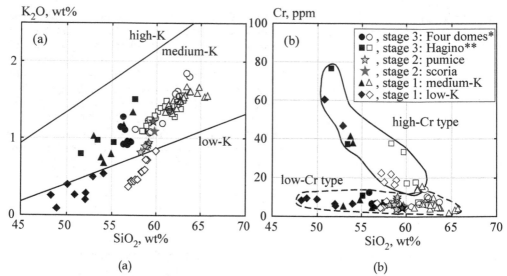

(a)   (b)

Fig. 2. K_2O vs. SiO_2 (a), in the left figure Cr vs. SiO_2 (b) in the right figure variation diagrams of the eruptive products from the Shirataka volcano (modified after Hirotani et al., 2009). Data are from Hirotani & Ban (2006). The boundary lines between low-K and medium-K, and medium-K and high-K are from Gill (1981). Filled symbols are mafic inclusions and open ones are host rocks. *Four domes consist of the Shiratakayama, Kitsunegoe, Nishikuromoriyama lava dome group + Higashikuromoriyama lava dome; **Hagino, the Hagino block-and-ash flow deposit

Petrologic Study of Explosive Pyroclastic Eruption Stage in Shirataka Volcano, NE Japan: Synchronized Eruption
of Multiple Magma Chambers

25

Furthermore, Hirotani et al. (2009) defined the high- and low-Cr types for the rocks from the Shirataka volcano. Cr and Ni contents of the former type are >20 ppm and >5 ppm, 48-55 wt% of SiO_2. Whereas, those of the latter type are ca. 15 ppm and ca. 3 ppm respectively for a given SiO_2 range of 48-55 wt%. Both types are observed in the low and medium-K series. The host lavas having high-Cr type mafic inclusions show higher Cr and Ni contents than those with low-Cr compositions. Consequently these lavas are referred to as high- and low-Cr type lavas, respectively (Fig. 2b). All the samples from stage 2 are low-Cr type.

In addition, the high-Cr rocks always have lower Sr isotope ratio than the low-Cr type rocks for a given range of silica contents of a same geologic or petrologic unit in the cases of stages 1 and 3 (Hirotani et al., 2009). All the samples from stage 2 show similar Sr isotope ratio to the medium-K, low-Cr type rocks in the first stage or those in the early part of the third stage.

4. Uranium, Thorium, and Hafnium contents of the eruptive products from the Shirataka volcano

Hirotani & Ban (2006) and Hirotani et al. (2009) pointed out that the compositions of the samples from the explosive pyroclastic eruption stage show similar features as the medium-K series, low-Cr type rocks of the other stages (Fig. 2), except for the following features. The pumices have extremely higher Zr contents than any other samples from the Shirataka volcano (Fig. 3), and these show slightly lower K_2O contents than the other medium-K lavas from the other stages (Fig. 2a). The higher Zr contents could be attributed to the presence of accessory minerals such as zircon and this would be confirmed by analysing Hf, Th, and U contents. (e.g., Rowe et al. 2007).

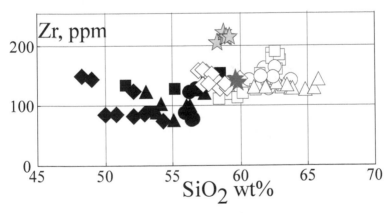

Fig. 3. Zr vs. SiO_2 diagram of the eruptive products from the Shirataka volcano (modified after Hirotani & Ban, 2006). Data are from Hirotani & Ban (2006). Symbols are the same as in Fig. 2.

4.1 Analytical method

Concentrations of Hf, Th and U were determined by ICP-MS on a VG Platform instrument at the Geological Survey of Japan/AIST following the method of Ishizuka et al. (2010). About 100 mg of sample powder was dissolved in a $HF-HNO_3$ mixture (5:1) using screw-top Teflon beakers on hotplate, followed by fusion with Na_2CO_3 (0.5 g) at 1,050°C. After evaporation to dryness, the residues were re-dissolved in 2% HNO_3 prior to analysis. In and

Re were used as internal standards, while JB2 and JB1a solutions diluted by the same method to the samples were used as external standards during ICP-MS measurements. Instrument calibration was performed using 5-6 solutions made from international rock standard materials. Reproducibility is generally better than ±6% (2 s.d.).

4.2 The analytical results

The representative data are listed in Table 2 and the data are plotted as the SiO_2 variation diagrams in Fig. 4.

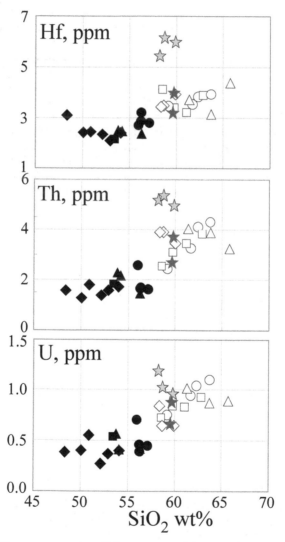

Fig. 4. Hf, Th, and U contents against SiO_2 contents of the samples from the Shirataka volcano. Symbols are the same as in Fig. 2.

Petrologic Study of Explosive Pyroclastic Eruption Stage in Shirataka Volcano, NE Japan: Synchronized Eruption
of Multiple Magma Chambers

27

stage	1	1	1	1	1	1
series	low-K	low-K	low-K	low-K	low-K	low-K
type	high-Cr	high-Cr	low-Cr	low-Cr	low-Cr	low-Cr
ppm						
Hf	2.12	2.45	3.93	2.36	3.50	2.44
Th	1.52	1.80	3.41	1.37	3.85	1.29
U	0.36	0.54	0.65	0.27	0.64	0.40
stage	1	1	1	1	1	1
series	low-K	low-K	medium-K	medium-K	medium-K	medium-K
type	low-Cr	low-Cr	high-Cr	high-Cr	high-Cr	low-Cr
ppm						
Hf	2.42	3.10	3.67	2.48	2.41	4.37
Th	1.69	1.52	3.95	2.23	2.12	3.15
U	0.40	0.39	1.00	0.56	0.40	0.87
stage	1	1	2 (pumice)	2 (pumice)	2 (pumice)	2 (scoria)
series	medium-K	medium-K	medium-K	medium-K	medium-K	medium-K
type	low-Cr	low-Cr	low-Cr	low-Cr	low-Cr	low-Cr
ppm						
Hf	3.12	2.33	6.16	5.39	5.96	3.23
Th	3.83	1.36	4.91	5.10	5.26	2.71
U	0.86	0.41	0.93	1.18	1.09	0.67
stage	2 (scoria)	3 (Hagino*)	3 (Hagino*)	3 (Hagino*)	3 (Hagino*)	3 (Hagino*)
series	medium-K	medium-K	medium-K	medium-K	medium-K	medium-K
type	low-Cr	high-Cr	high-Cr	high-Cr	low-Cr	low-Cr
ppm						
Hf	4.02	3.22	2.17	3.40	3.90	4.12
Th	3.70	3.41	1.83	3.07	3.82	2.52
U	0.89	0.83	0.54	0.84	0.92	0.72
stage	3 (Domes**)	3 (Domes**)	3 (Domes**)	3 (Domes**)	3 (Domes**)	3 (Domes**)
series	medium-K	medium-K	medium-K	medium-K	medium-K	medium-K
type	low-Cr	low-Cr	low-Cr	low-Cr	low-Cr	low-Cr
ppm						
Hf	3.84	2.90	3.47	2.82	3.93	2.72
Th	4.05	1.63	2.44	1.59	4.25	2.54
U	1.03	0.39	0.75	0.45	1.10	0.70
stage	3 (Domes**)	3 (Domes**)				
series	medium-K	medium-K				
type	low-Cr	low-Cr				
ppm						
Hf	3.54	3.21				
Th	3.23	1.64				
U	0.94	0.45				

Table 2. Hf, Th, and U contents of representative samples from the Shirataka volcano.
Hagino*, the Hagino block-and-ash flow deposit; Domes**, the Shiratakayama, Kitsunegoe,
Nishikuromoriyama lava dome group + Higashikuromoriyama lava dome

The Hf, Th, and U contents of rocks from stages 1 and 3 are plotted on the same compositional trends in the silica variation diagrams. These contents gradually increase with increasing SiO_2 content (Fig. 4). The ranges of Hf, Th, and U contents of rocks from stages 1 and 3 are 2.1 to 4.4, 1.4 to 4.1, and 0.3 to 1.0, respectively (Table 2). The ranges of Hf, Th, and U contents of scoria samples from stage 2 fall in the ranges of those from stages 1 and 3. However, the pumice samples of stage 2 are plotted in the higher areas than the compositional trends defined by the samples from stages 1 and 3 in the Hf, Th, and U vs. SiO_2 diagrams (Fig. 4). The contents of these elements in the pumice samples are extremely higher than the other ones (Table 2).

5. Discussion

Previous petrologic studies (Hirotani & Ban, 2006; Hirotani et al., 2009) on samples from the first and third stages revealed that (1) all products were formed by magma mixing between mafic and felsic end-member magmas, (2) both end-member magmas changed their compositions with time, (3) the felsic end-member magma was formed by the melt extraction or re-melting of partially or fully crystallized mafic end-member magma of the same activity. The evidence for (1) includes the existence of mafic inclusions in all lava samples, disequilibrium phenocryst assemblages (Sakuyama, 1981) such as Mg-rich olivine and quartz, and wide compositional ranges in plagioclase phenocryst in all samples. The estimated end-member magmas for geologic or petrologic units of stages 1 and 3 are presented in Table 3. As can be observed in the table, the compositions of both the two end-members change temporally.

stage 3: lava-dome & block-and-ash flow forming stage
Four lava domes*   ; medium-K, low-Cr type ; 52.00% SiO_2 and 0.60% K_2O (mafic end-member),
                                                    70.00% SiO_2 and 2.20% K_2O (felsic end-member)
Hagino block-and-ash ; medium-K, high-Cr type ; 49.50% SiO_2 and 0.64% K_2O (mafic end-member),
flow deposit   66.50% SiO_2 and 1.75% K_2O (felsic end-member)
Hagino block-and-ash ; medium-K, low-Cr type ; 50.50% SiO_2 and 0.80% K_2O (mafic end-member),
flow deposit   66.00% SiO_2 and 1.65% K_2O (felsic end-member)

stage 1: strato-cone building stage
Kokuzo lava   ; medium-K, high-Cr type ; 51.00% SiO_2 and 0.50% K_2O (mafic end-member),
                                                    66.50% SiO_2 and 1.85% K_2O (felsic end-member)
Kokuzo lava   ; medium-K, low-Cr type ; 52.00% SiO_2 and 0.85% K_2O (mafic end-member),
                                                    67.50% SiO_2 and 1.70% K_2O (felsic end-member)
Kokuzo lava   ; low-K, high-Cr type   ; 48.00% SiO_2 and 0.20% K_2O (mafic end-member),
                                                    64.50% SiO_2 and 0.85% K_2O (felsic end-member)
Kokuzo lava   ; low-K, low-Cr type   ; 49.00% SiO_2 and 0.21% K_2O (mafic end-member),
                                                    64.00% SiO_2 and 0.75% K_2O (felsic end-member)

Table 3. Temporal change of the mafic and felsic end-members compositions from stages 1 and 3 in the Shirataka volcano. Data are from Hirotani et al. (2009). *Four lava domes, the Shiratakayama, Kitsunegoe, Nishikuromoriyama lava domes group + Higashikuromoriyama lava dome

Similarity of Sr and Nd isotope ratios within the same Cr-type host rocks and mafic inclusions of the same geologic unit shows the consanguinity of the mafic and felsic end-members for each case. The trace element calculations showed that the felsic end-member

Petrologic Study of Explosive Pyroclastic Eruption Stage in Shirataka Volcano, NE Japan: Synchronized Eruption
of Multiple Magma Chambers

29

cannot be explained by the fractional crystallization process but by the remelting process of the solidified mafic end-member leaving behind gabbroic residue (Hirotani et al., 2009). The genesis of felsic magmas have been explained by similar process in many cases (e.g. Feeley et al, 1998; Hansen et al., 2002, Smith et al., 2006; Vogel et al, 2006; Ban et al., 2007).

The pumice and scoria in the stage 2 possess evidence of mama mixing, such as dissolution textures in plagioclase phenocrysts, disequilibrium phenocryst assemblages, and wide compositional variation in plagioclase phenocrysts. Thus the pumice and scoria could also be formed by magma mixing (Hirotani & Ban, 2006). The compositions of the end-member magmas for the products of stages 1 and 3 have been well discussed (Hirotani et al., 2009), because the compositional ranges of the samples from stages 1 and 3 are wide and thus it was possible to determine the compositions of end-member magmas using the magma mixing lines in the compositional variation diagrams. However, pumice and scoria samples from the stage 2 display very narrow compositional ranges. It was impossible to determine the end-member compositions in these cases using the same method applied for stages 1 and 3 samples. Thus the compositions of the end-member magmas still remain in unknown. In the following sections, we will define these end-member compositions using another method.

5.1 Estimation of the compositions of the end-member magmas for the pumice and scoria in the explosive pyroclastic eruption stage

The juvenile fragments of the explosive pyroclastic eruption stage are mainly pumice with minor amount of scoria, and these also possess petrologic features of mixing of two end-members (Hirotani & Ban, 2006). However there is very little clue to estimate the end-member compositions because of the very narrow compositional ranges of the pumice and the scoria. Thus, here we assume that the end-member compositions of the pumice and scoria are similar to one of the estimated end-member compositions for stages 1 and 3 eruptive products. The chemical compositions of the scoria fall in the ranges of the products of the early part of the third stage (Figs. 2, 3 & 4). Thus the end-member compositions of the scoria would be the same as in the early part of the third stage. On the other hand, the pumice shows higher Zr, Hf, Th, and U contents than any other samples in the Shirataka volcano.

In Fig. 5, we compare the compositions of glass inclusions in pyroxene phenocrysts of pumice to those from stages 1 and 3. Representative compositional data of the glass inclusions are presented in Hirotani & Ban (2006). The compositional range of the glass compositions from pumice is similar to those from the medium-K products of stage 1 in Fig. 5. Therefore, it is reasonable to consider that the felsic end-member compositions of the pumice are also similar to the above mentioned ones. In this case, the mafic end-member compositions should plot on the silica poorer extensions of the mixing lines tying the felsic end-member and pumice compositions (Fig. 6). In this figure, the mafic end-member of the low-K samples of stage 1 is on the extension line (orange colored dotted line in Fig. 6). Here, we assume that the compositions of the mafic end-member are the same as those of low-K products of stage 1. In the next step, we determined the felsic end-member compositions of the pumice, using the tie lines of the assumed mafic end-member and pumice compositions on the SiO_2 variation diagrams, assuming the silica content of the felsic end-member is the same as that of the medium-K products of stage 1. The obtained chemical compositions of the felsic end-member are similar to those of the medium-K products of stage 1, except for higher Zr, Hf, Th, and U contents.

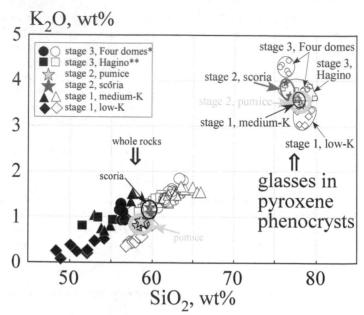

Fig. 5. K₂O-SiO₂ diagram showing the compositions of glass inclusions in pyroxene phenocrysts from each stage in the Shirataka volcano. *Four domes, the Shiratakayama, Kitsunegoe, Nishikuromoriyama lava dome group + Higashikuromoriyama lava dome; **Hagino, the Hagino block-and-ash flow deposit

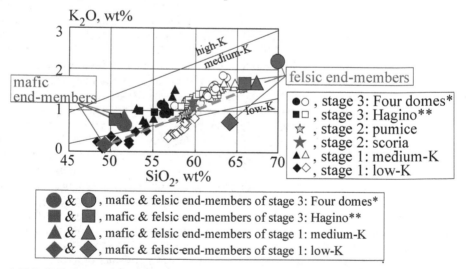

Fig. 6. K₂O-SiO₂ diagram showing the end-member compositions for the pumice and scoria from stage 2 in the Shirataka volcano. The boundary lines of low-K and medium-K, and medium-K and high-K are from Gill (1981). The dotted line shows the mixing line for the pumice. *Four domes, the Shiratakayama, Kitsunegoe, Nishikuromoriyama lava dome group + Higashikuromoriyama lava dome; **Hagino, the Hagino block-and-ash flow deposit

Petrologic Study of Explosive Pyroclastic Eruption Stage in Shirataka Volcano, NE Japan: Synchronized Eruption
of Multiple Magma Chambers

31

5.2 Melting of accessory zircon with thorite inclusions in producing the felsic end-member magmas for the pumice in stage 3 in the Shirataka volcano

During dormancy between stages 1 and 2, the magma chamber filled with the felsic end-member of the medium-K samples in the first stage would become a mush chamber (e.g., Huber et al., 2010) by the crystallization. In the second stage, new magma injection from deeper levels would re-activate the mushy chamber. This re-activated magma would become to felsic end-member magma for the pumice in the second stage. Such re-activation process of the old magma chambers is thought to be common in producing felsic magmas (e.g., Tamura & Tatsumi, 2002; Girard & Stix, 2009; McCurry & Rodgers, 2009). The higher Zr, Hf, Th, and U contents of the felsic end-member magma can be explained by the melting of the accessory zircon minerals during the re-activation (Wolff et al., 2006). The estimated Zr content of the felsic magma can be obtained by melting of zircon-bearing rock (0.04% of the melt should be from zircon melting). Accordingly, the Hf, Th, and U contents in the zircon are calculated to be ca. 11,500, 10,000, and 1,100 ppm, respectively (Fig. 7). The Hf, Th, and U contents in zircon minerals were reported by many researches (e.g., Rubatto & Hermann, 2007). The calculated Hf content of ca. 11,500 ppm is in the range of reported Hf contents in zircon. On the other hand, the calculated Th and U contents of ca. 10,000 and 1,100 ppm are exceed the ranges of Th and U contents in zircon in equilibrium with the felsic end-member (e.g., Rubatto & Hermann, 2007). Recently, Paquette & Mergoil-Daniel (2009) reported that zircon crystals sometimes include tiny thorite inclusions. The Th and U contents in thorite are much higher than in zircon. Thus, the high Th and U contents can be ascribed to the melting of thorite-bearing components during the re-activation.

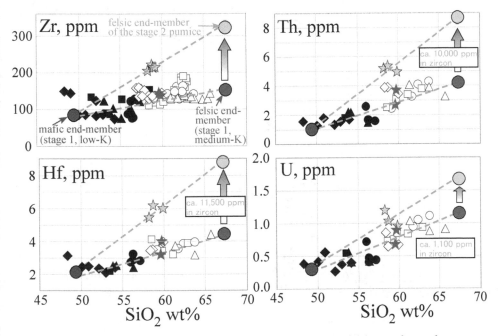

Fig. 7. Plots of Zr, Hf, Th, and U vs. SiO_2 showing the estimated felsic end-member magma compositions for the studied pumice. Symbols, other than the end-members are the same as in Fig. 2.

5.3 The magma feeding system for the explosive pyroclastic eruption stage
5.3.1 Formation of mixed magma for the pumice

As discussed above, the felsic end-member of medium-K samples of the first stage would be re-activated at the second stage and became the felsic end-member magma for the pumice. The end-member magma chamber would be stored in shallower part of the crust. The re-activation would be induced as a result of the ascent of the mafic end-member of the low-K products of the first stage. This magma would be still active also in the second stage (Fig. 8). The mafic magma from deeper levels is characterized by high-Cr type. This type mafic magma differentiate into low-Cr type through the AFC process as described for the formation of the low-Cr mafic magmas of stages 1 and 3 by Hirotani et al. (2009). The low-Cr type mafic magma would ascent further. When the mafic magma reached shallow crustal level where the remnant of stage 1 medium-K felsic magma chamber locates, the mafic magma would re-activate this felsic chamber by its heat. The higher Zr, Hf, U, and Th contents of the felsic end-member magma can be explained by the melting of the accessory zircon minerals with thorite inclusions during the re-activation. The re-activated felsic magma would mix with the low-Cr type mafic magma in the chamber. The schematic representation of the magma feeding system is presented in Fig. 8.

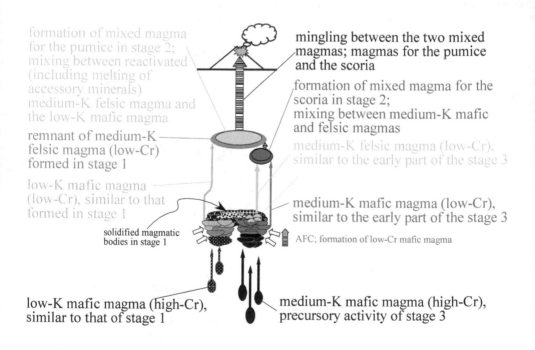

formation of mixed magma for the pumice in stage 2; mixing between reactivated (including melting of accessory minerals) medium-K felsic magma and the low-K mafic magma

mingling between the two mixed magmas; magmas for the pumice and the scoria

remnant of medium-K felsic magma (low-Cr) formed in stage 1

formation of mixed magma for the scoria in stage 2; mixing between medium-K mafic and felsic magmas

low-K mafic magma (low-Cr), similar to that formed in stage 1

medium-K felsic magma (low-Cr), similar to the early part of the stage 3

solidified magmatic bodies in stage 1

medium-K mafic magma (low-Cr), similar to the early part of the stage 3

AFC; formation of low-Cr mafic magma

low-K mafic magma (high-Cr), similar to that of stage 1

medium-K mafic magma (high-Cr), precursory activity of stage 3

Fig. 8. Schematic representation of the magma feeding system of the stage 2 in the Shirataka volcano.

Petrologic Study of Explosive Pyroclastic Eruption Stage in Shirataka Volcano, NE Japan: Synchronized Eruption of Multiple Magma Chambers

33

5.3.2 Formation of mixed magma for the scoria

The third stage mafic end-member magma would antecedently appear as the mafic end-member for scoria at the second stage. The felsic end-member magma for the scoria was generated by the melt extraction or re-melting process, and subsequently mixed with the mafic end-member in a crustal magma chamber. The Mg# and the estimated precipitated temperature of two-pyroxenes in the scoria are higher than those in the other felsic end-members (Hirotani & Ban, 2006). Furthermore the size of the pyroxene phenocrysts is smaller than those in the other felsic end-members, but larger than the groundmass minerals. These features indicate that the pyroxenes of the scoria are grouped to group III (population III of Hirotani et al. (2009)), and precipitated from the mixed magma for the scoria in the chamber, suggesting that the mixed magma was stored in the crust for a while.

5.3.3 Mingling between two mixed magmas

Consequently, two kinds of mixed magma were stored at shallow crust level. One is for the pumice and the other is for the scoria. Further ascent of the mafic end-member triggered the ascent of the mixed magma of the scoria. This mixed magma would tap the shallow chamber filled with the other mixed magma of the pumice. Finally, these two mixed magmas erupted synchronously, which made the eruption explosive. In addition, rarely observed banded pumice would be formed probably during the ascent through the conduit.

6. Conclusions

Detailed petrologic study on the explosive pyroclastic eruption stage of the Shirataka volcano, NE Japan, produced the following results. The eruptive products are mostly pumice with minor amount of scoria. Banded pumice is rarely observed.

1. Both the pumice and scoria were formed by two end-members mixing. The estimated end-member components for the pumice are similar to those activated in the first stage. Whereas, those for the scoria are similar to those of the third stage.
2. The felsic end-member magma for the pumice shows extremely high Zr, Hf, Th, and U contents than felsic magmas in the other stages. These high contents can be explained by the melting of accessory zircon crystals, which have thorite inclusions, when the felsic magma was formed by the re-activation of previously stalled felsic chamber.
3. The hybrid magmas for the pumice and scoria were formed in the crustal chambers before the explosive eruption. During the synchronized eruption, the mixing/mingling between the two magmas occurred, resulting in the formation of the banded pumice.
4. The second stage was transitional one from the first to the third, and the mafic magmas of the first and third stages were simultaneously ascended to the shallow magma feeding system. The former re-activated the felsic magma chamber formed in the first stage, while the latter is regarded as the antecedent activity of the third stage. Finally, there was the ascent of the mixed magma from depth that subsequently tapped the shallow chamber filled with the other mixed magma, which caused the explosive eruption.

7. Acknowledgment

We greatly appreciate anonymous reviewer for many constructive review comments on the early version of the manuscript. The manuscript was greatly improved by the comments. We are grateful for Dr. S. Nakano for supporting this study. We also thank K. Yamanobe for assisting with the ICP-MS measurements.

8. References

Bacon, C. R. (1986) Magmatic inclusions in silicic and intermediate volcanic rocks. *J. Geophys. Res,* 91, 6091-6112.

Ban, M., Hirotani, S., Wako, A., Suga, T., Iai, Y., Kagashima, S., Shuto, K. & Kagami, H. (2007). Origin of silicic magmas in a large-caldera-related stratovolcano in the central part of NE Japan - Petrogenesis of the Takamatsu volcano -. *J. Volcanol. Geotherm. Res.,* 167, 100-118.

Eichelberger, J. C. (1980) Vesiculation of mafic magma during replenishment of silicic magma reservoirs. *Nature,* 288, 446-450.

Feeley, T. C., Dungan, M. A. & Frey, F. A. (1998) Geochemical constraints on the origin of mafic and silicic magmas at Cordon El Guadal, Tatara-San Pedro Complex, central Chile. *Contrib. Mineral. Petrol.,* 131, 393-411.

Gill, J. B. (1981) *Orogenic andesites and plate tectonics,* Springer-Verlag, Berlin. pp 385.

Girard, G. & Stix, J. (2009) Magma recharge and crystal mush rejuvenation associated with early post-collapse upper basin member rhyolites, Yellowstone Caldera, Wyoming. *J. Petrol.,* 50, 2095-2125.

Hansen, J., Skjerlie, K. P., Pederson, R. B. & Rosa, J. D. L. (2002) Crustal melting in the lower parts of island arcs: an example from the Bremanger Granitoid Complex, west Norwegian Caledonides. *Contrib. Mineral. Petrol.,* 143, 316-335.

Hirotani, S. & Ban, M. (2006) Origin of silicic magma and magma feeding system of Shirataka volcano, NE Japan. *J. Volcanol. Geotherm. Res.,* 156, 229-251.

Hirotani, S., Ban, M. & Nakagawa, M. (2009) Petrogenesis of mafic and associated silicic end-member magmas for calc-alkaline mixed rocks in the Shirataka volcano, NE Japan. *Contrib. Mineral. Petrol.,* 157, 709-734.

Huber, C., Bachmann, O. & Dufek, J. (2010) The limitations of melting on the reactivation of silicic mushes. *J. Volcanol. Geotherm. Res.,* 195, 97-105.

Ishii, M. & Saito, K. (1997) A K-Ar age study on Shirataka volcano, Yamagata Prefecture. *Bull. Yamagata Univ., Nat. Sci.,* 14, 99-108.

Ishizuka, O., Yuasa, M., Tamura, Y., Shukuno, H., Stern, R. J., Naka, J., Joshima, M. & Taylor, R. N. (2010) Migrating shoshonitic magmatism tracks Izu–Bonin–Mariana intra-oceanic arc rift propagation. *Earth Planet. Sci. Lett.,* 294, 111–122.

Kawano, Y., Yagi, K. & Aoki, K. (1961) Petrography and petrochemistry of the volcanic rocks of Quaternary volcanoes of northeastern Japan. *Sci. Rep. Tohoku Univ. Ser III,* 7, 1-46.

Mimura, K. & Kanno, K. (2000) Stratigraphy and history of Shirataka volcano, NE Japan. *Bull. Volcanol. Soc. Jpn.,* 45, 13-23.

McCurry, M. & Rodgers, D. W. (2009) Mass transfer along the Yellowstone hotspot track I: petrologic constraints on the volume of mantle-derived magma. *J. Volcanol. Geotherm. Res.*, 188, 86-98.

Nakagawa, M., Shimotori, H. & Yoshida, T. (1988) Across-arc compositional variation of the Quaternary basaltic rocks from the Northeast Japan arc. *J. Mineral. Petrol. Econ. Geol.*, 83, 9-25.

Nakagawa, M., Kitagawa, J. & Furukawa, R. (2003) Sequential caldera-forming eruptions from multiple magma chambers of Shikotsu caldera, Hokkaido, Japan. *XXIII General Assembly of the International Union of Geodesy and Geophysics*, Abstract V08/02P/A02-005.

Paquett, J. L. & Mergoil-Daniel, J. (2009) Origin and U–Pb dating of zircon-bearing nepheline syenite xenoliths preserved in basaltic tephra (Massif Central, France). *Contrib. Mineral. Petrol.*, 158, 245–262.

Rowe, M. C., Wolff, J. A., Gardner, J. N., Ramos, F. C., Teasdale, R. & Heikoop, C. E. (2007) Development of a continental volcanic field: petrogenesis of pre-caldera intermediate and silicic rocks and origin of the Bandelier magmas, Jemez Mountains (New Mexico, USA). *J. Petrol.*, 48, 2063-2091.

Rubatto, D. & Hermann, J. (2007) Experimental zircon/melt and zircon/garnet trace element partitioning and implications for the geochronology of crustal rocks. *Chem. Geol.*, 241, 38–61.

Sakuyama, M. (1981) Petrological study of the Myoko and Kurohime volcanoes, Japan: crystallization sequence and evidence for magma mixing. *J. Petrol.*, 22, 553-583.

Shane, P., Martin, S. B., Smith, V. C., Beggs, K. F., Darragh, Cole, J. W. & Nairn, I. A. (2007) Multiple rhyolite magmas and basalt injection in the 17.7 ka Rerehakaaitu eruption episode from Tarawera volcanic complex, New Zealand. *J. Volcanol. Geotherm. Res.*, 164, 1-26.

Smith, I. S. E., Worthington, T. J., Price, R. C., Stewart, R. B. & Maas, R. (2006) Petrogenesis of dacite in an oceanic subduction environment: Raoul Island, Kermadec arc. *J. Volcanol. Geotherm. Res.*, 156, 252-265.

Tamura, Y. & Tatsumi, Y. (2002) Remelting of an andesitic crust as a possible origin for rhyolitic magma in oceanic arcs: an example from the Izu-Bonin arc. *J. Petrol.*, 43, 1029-1047.

Tatsumi, Y. & Eggins, S. (1995) *Subduction zone magmatism*, Blackwell: Oxford, pp 211.

Tsuchiyama, A. (1985) Dissolution kinetics of plagioclase in the melt of the system diopside-albite-anorthite, and origin of dusty plagioclase in andesites. *Contrib. Mineral. Petrol.*, 89, 1-16.

Wolff, J. A., Wark, D. A., Ramos, F. C. & Olin, P. H. (2006) Petrologic evidence for thermal rejuvenation of crystal mush in the Bandelier Tuff. *Eos Trans. AGU, 87(52), Fall Meet. Suppl.*, Abstract V24C-01.

Vogel, T. A., Patino. L. C., Jonathon K. Eaton, J. K., Valley, J. W., Rose, W. I., Alvarado, G. E. & Viray, E. L. (2006) Origin of silicic magmas along the central American volcanic front: Genetic relationship to mafic melts. *J. Volcanol. Geotherm. Res.*, 156, 217-228.

Yagi, H., Soda, T, Inokuchi, T., Haraguchi, T. & Ban, M. (2005) Catastrophic collapse of Mt. Zao and Mt. Shirataka and their chronological timing. *The Quaternay Res.*, 44, 263-272.

3

Petrological and Geochemical Characteristics of Mafic Granulites Associated with Alkaline Rocks in the Pan-African Dahomeyide Suture Zone, Southeastern Ghana

Prosper M. Nude[1], Kodjopa Attoh[2],
John W. Shervais[3] and Gordon Foli[4]

[1]Department of Earth Science, University of Ghana, P.O. Box LG 58, Legon-Accra,
[2]Department of Earth &Atmospheric Sciences, Cornell University, Ithaca, NY 14853,
[3]Department of Geology, Utah State University, Logan UT 84322,
[4]Department of Earth and Environmental sciences, University for
Development studies, Navrongo Campus
[1,4]Ghana
[2,3]USA

1. Introduction

Most alkaline complexes are characterized by the presence of a distinctive zone where alkaline emanations appear to affect the wall rocks and the contact zones with country rocks (Winter, 2001). Such alkaline solutions and magmas may be effective agents for transporting trace elements and modifying the compositions of the host rocks (Wallace & Green, 1988, Rudnick et al., 1993). As a result the primary minerals can be replaced by alkaline minerals such as nepheline and feldspar. In this way nepheline-bearing rocks and other metasomatic derivatives of variable compositions can arise (Dawson et al. 1990). In the Pan-African Dahomeyide suture zone in southeastern Ghana, variably deformed alkaline rocks, comprising nepheline syenite and carbonatitic rocks, referred to as the Kpong complex (KC), occur in tectonic contact with high-pressure (HP) mafic granulite rocks of garnet-pyroxene-amphibole composition (Nude et al., 2009). The Dahomeyide mafic granulites have been found to preserve geochemical imprints of island arc theoleiitic (IAT) basalts as well as rocks with N-MORB-like affinities (Agbossoumonde et al., 2001, Attoh & Morgan 2004). Thus the mafic granulites possess distinct geochemical signatures that differ significantly from the alkaline rocks.

This paper presents petrological and geochemical data on the nepheline-bearing mafic rocks previously referred to as mafic nepheline gneiss (Holm, 1974) at the contact zone between the HP mafic granulites and the KC rocks. The data are used to evaluate the distinctive mineralogical and trace element contents of the nepheline-bearing mafic rocks, and also infer the interactions of the alkaline magma with the mafic granulites at the contact zone.

2. Regional geological setting

The Dahomeyide orogen in southeastern Ghana and adjoining parts of Togo and Benin is the southern segment of the Pan-African Trans-Saharan belt (TSB). The TSB defines the eastern margin of the West African craton (WAC) and extends for over 2500 km from the Sahara to the Gulf of Guinea (Caby, 1987). The Pan-African orogen resulted in the assembly of northwest Gondwana (Hoffman, 1991; Cordani et al., 2003; Tohver et al., 2006). In southeastern Ghana and adjoining parts of Togo and Benin the Dahomeyide is interpreted to have resulted from easterly subduction after resorbtion of oceanic lithosphere at rifted margin of WAC (Affatton et al., 1991; Agbosoumonde et al., 2004, Attoh & Nude, 2008) with a preserved suture. These rocks are also exposed in the Amalaoulaou complex to the north in the Gourma fold and thrust belt in Mali (Berger et al., 2011) and shares comparable geochemical, metamorphic and tectonic evolution to the rocks of the Dahomeyides to the south in Benin, Togo and Ghana.

Figure1 is a geologic map of the Dahomeyide orogen in southeastern Ghana, and adjoining parts of southern Togo and Benin (Sylvain et al., 1986, Castaing et al., 1993; Attoh et al.,

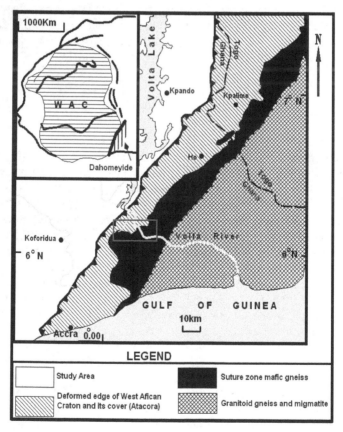

Fig. 1. Tectonic map of the Dahomeyides in southeastern Ghana and its northern extension (After Attoh, 1998) showing the study area

Petrological and Geochemical Characteristics of Mafic Granulites Associated with Alkaline Rocks
in the Pan-African Dahomeyide Suture Zone, Southeastern Ghana

39

1997) showing the principal lithologies of the orogen. From the west is the deformed margin of the WAC that include 2.1 Ga granitoids (Agyei et al., 1987; Agbossoumonde et al., 2007), known as Ho gneisses, now deformed into proto-mylonites, and its cover rocks (Atacora nappes) occurring on the rifted passive margin. These are bounded to the east by distinctive high-pressure (HP) mafic granulite and eclogite facies rocks known locally as the Shai-Hill gneisses that form the suture zone unit (Attoh, 1998; Agbossoumonde et al., 2001; Attoh & Morgan, 2004) and mark the zone of collision of WAC with presumed exotic blocks to the east. Granitoids to the east of the suture zone comprise migmatites and dioritic gneisses which represent the arc terrane that is postulated to have formed during the subduction and accompanying oceanic closure.

3. Lithological distributions and previous geochronological work

3.1 Lithological distribution

The lithological distributions of the alkaline rocks in relation to the mafic granulite gneiss and other lithological units have been described by several workers including Holm (1974), Attoh et al. (2007), Nude et al. (2009), and the geology is shown in Figure 2. The alkaline rocks comprise alternating layers and interfolded units of nepheline syenite gneiss and carbonatite along the inferred sole thrust of the suture that separates the mafic granulite

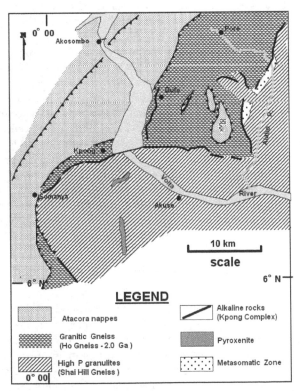

Fig. 2. Geological map of the study area showing the lithological relatioships and the metasomatic zone where the samples were taken.

gneiss from rocks of the deformed edge of the WAC. The nepheline-bearing mafic granulite which forms the basis of this study is a garnet-bearing rock that is restricted to the contact zone with the alkaline rocks and the Shai Hills gneisses. It occurs in isolated outcrops in the northeast of the area (Fig. 2) where it is typically folded with steep axial surfaces, subvertical hinge zones and asymmetrical limbs. Attoh et al. (1997) interpreted the structure of the suture zone to have resulted from early east–west compression, which produced the north–south imbricate thrust slices followed by NNW-directed thrusting.

3.2 Previous geochronological work

Geochronological studies of the suture zone mafic granulite gneisses (Shai Hill gneisses) and the alkaline rocks provide constraints on the chronology of the tectonic record of the area. U-Pb zircon ages determined from the mafic granulites from the suture zone in Ghana by Attoh et al. (1991) and interpreted as peak metamorphic age was 610 ± 2 Ma. Also Hirdes and Davis (2002) reported U-Pb zircon ages of 603 ± 5 Ma from the mafic granulites from the Shai Hills area which confirm the timing of peak metamorphism in the suture zone. Similar age of 613 ± 1 Ma from zircon evaporation ($^{207}Pb/^{206}Pb$) was reported by Affaton et al. (2000) for the suture zone rocks in northern Togo. Hornblende separates from the mafic granulites yielded $^{40}Ar/^{39}Ar$ ages between 587 and 567 Ma, interpreted as the time of exhumation of the nappes (Attoh et al., 1997). Thus taken together high pressure metamorphism of the suture zone rocks occurred around 603-613 Ma and exhumation through the hornblende ages around 580-570 Ma (Attoh et al., 2007). U-Pb ages on zircon separates determined by Bernard-Grifiths et al. (1991) from eclogite facies rocks from the suture zone in southern Togo had a discordant lower intercept of ~640 ± 53 Ma and Nd model ages (T_{DM}) of 1150 Ma (Bernard-Grifiths et al., 1991). Nd model age of 940 Ma was obtained by Attoh and Schmitz (1991) in the HP mafic granulites from the Shai Hills area in Ghana. The model ages suggest that the mantle derivation of the protoliths of these rocks may have occurred earlier. In the Amalaoulaou arc in Mali, the magmatic activity was found to have occurred at least c. 793 - 660 Ma followed by UHP metamorphism at c. 623 Ma (Berger et al., 2011).

Analyses of zircons separates from the carbonatite and the nepheline syenite gneiss samples yielded ages of 592-594 ± 4 Ma interpreted as the time of intrusion of the alkaline rocks (Nude et al., 2006). So the available age data suggest the emplacement of the alkaline rocks during syn-orogenic rifting, but this occurred after peak granulite metamorphism (Attoh et al., 2007). Overall therefore the alkaline rocks appear to have been emplaced later than the mafic granulites.

4. Petrographic and geochemical characteristics of the mafic granulites in the suture zone

The mafic granulites (Shai Hills gneiss) are variably sheared and deformed, and have a streaky appearance. The rocks are characterized by prominent modal layering consisting of alternating but discontinuous garnet-rich and hornblende—rich zones that are cut by veins of all sizes and orientations. The microstructural features of the Shai Hills gneisses have been described by Attoh and Nude (2008). Generally the rocks are composed of variable proportions of garnet, diopside pyroxene and scapolite. The following petrographic types have been identified by Attoh (1998): a) hornblende-rich granulite with typical modal compositions of 42% hornblende, 38 % plagioclase, 9 % garnet, 4% diopside and 5% quartz,

Petrological and Geochemical Characteristics of Mafic Granulites Associated with Alkaline Rocks
in the Pan-African Dahomeyide Suture Zone, Southeastern Ghana

41

and b) garnet-rich granulites that have similar mineral assemblage but with different mineral proportions of 29% garnet, 26% plagioclase, 20% diopside, 9% hornblende, 10% quartz and 2% scapolite. Geochemical features determined by Attoh and Morgan (2004) suggest that the mafic granulites have predominantly island arc tholeiite imprints with subordinate N-MORB signatures and trace element patterns that are very similar to lower crust compositions.

5. Petrographic and geochemical characteristics of the alkaline rocks

The alkaline rocks consist of nepheline syenite gneiss and carbonatite, and their petrographic features have been described by Holm (1974), Nude et al. (2009). The nepheline syenite gneiss is composed of nepheline (20–30%) which sometimes shows replacement by cancrinite, Other major phases are sodic feldspar (An0–An4, 30–50%), perthitic microcline and/or orthoclase (15–30%), annitic biotite (5–15%). Titanite is a widespread accessory constituent. Minor accessories include fine grained calcite, zircon, apatite, and muscovite. More syenitic varieties occur locally consisting essentially of albite, microcline, accessory biotite and nepheline. Modally, the carbonatite consists of coarse-grained mosaics of subhedral to euhedral equant calcite (35–50%) and annitic biotite (25–40%), with feldspar (albite and microcline/orthoclase, 5–20%) and nepheline (2–20%) and rare zircon.

Common mineral phases such as calcite, nepheline, feldspar and biotite in the nepheline syenite gneiss and the carbonatite have similar compositions (Attoh & Nude, 2008; Nude et al., 2009). The calcites show homogeneous compositions; CaO concentrations fall within 49.07–57.36 wt% and they are enriched in Sr with SrO values up to 1.4 wt%. Nepheline in both rock suites is generally similar in composition; it is relatively sodium rich, and compositions fall within $Na_{6.0-8.1}K_{0.4-1.7}Al_{7.3-7.9}Si_{8.0-8.2}O_{32}$. K-feldspar in the rocks is almost pure orthoclase with over 94 mol% Or in the nepheline syenite. Plagioclase is essentially albite, and common in almost all samples with compositions from 78 to 99 mol% Ab in the carbonatite, 94–98 mol% Ab in some nepheline syenite gneiss samples, confirming the compositional similarities in both rock suites. Biotite from the rocks is generally annitic with the composition falling within $K_{1.8-1.9}Fe_{3.1-3.5}Mg_{1.2-1.4}Si_{5.2-5.3}Al_{3.1-3.4}O_{20}(OH,F_{0.1-0.4})$. Geochemically the alkaline rocks are characteristically enriched in alkalis ($Na_2O + K_2O$ is up to 16.4 wt %), Ba (3389-4665 ppm), Sr (3891-5481 ppm), Nb (78-135 ppm). The rocks show strong LREE fractionations and large deletions of Zr and Hf relative to primitive mantle (Nude et al., 2009). Most carbonatite and related rocks worldwide are known to have these geochemical features (Potter, 1996; Nelson et al., 1988; Woolley & Kemp, 1989; Hornig-Kjarsgaard, 1998; Bell & Tilton, 2001; Thompson et al., 2002; Chakhmouradian et al., 2007).

6. Petrography of the mafic granulites in the metasomatic zone

Representative samples of the mafic granulites analyzed in this study were taken from the metasomatic zone (Fig. 2). Generally these rocks which were previously mapped as mafic nepheline gneiss (Holm, 1974, Kesse, 1985) are found in isolated outcrops as a dense, foliated rock close to the alkaline rocks. The dark colour, coarse texture and significant modal content of garnet and pyriboles make the mafic granulite gneiss conspicuous in the bluish-gray nepheline gneiss and the dark-grey carbonatite. The rock contains feldspar and nepheline rich veinlets in the shear zone. Major modal compositions are variable and are composed of garnet (10-25 vol. %), sodic plagioclase (~30 vol. %), microcline (~15 vol. %), nepheline (~20 vol. %), aegirine–augite (~35 vol. %), ferro-pargastite amphibole (10-30 vol. %), coarse titanite (~5 vol. %). The feldspars are generally coarse but in some of the crystals they occur as equigranular, granoblastic and interstitial grains. Accessory constituents include calcite, mostly found in cleavage cracks, zircon and rare kaersutite.

6.1 Composition of common mineral phases in the mafic granulites from the metasomatic zone and the alkaline rocks

The common mineral phases in the mafic granulites from the metasomatic zone and the alkaline rocks are calcite, nepheline, and feldspar. The compositions of these mineral phases were determined from representative samples of the mafic granulites with the objective of comparing their chemical contents with those from the alkaline rocks determined from previous studies by Attoh and Nude (2008) and then Nude et al. (2009). This will provide an insight into the extent of similarities in these common phases in the adjacent rocks. Two representative samples PN32A and PN56 which represent the variability of the compositional phases were selected for phase chemistry analysis. The mineral chemistry analysis was done using a Cameca SX-50 electron microprobe at the University of Utah. The minerals were tentatively identified using energy dispersive spectrometry (EDS). Table 1 lists the results of the microprobe analysis.

6.1.1 Calcite

Calcite is the only carbonate in the rocks; CaO contents range from 51.0 – 53.8 wt %. The totals of the major element concentrations are limited and fall within 55-58 wt % excluding volatiles and. The mineral is characteristically Sr-rich, with values within 1.3- 1.5 wt %.

6.1.2 Nepheline

Nepheline compositions in the rocks are variable, but a key feature is that it is Na-rich, and the variable compositions fall within $Na_{2.9-6.0}K_{0.0-1.7}Al_{4.2-8.2}Si_{8.0-11.8}O_{32}$. Two varieties of the nepheline have been recognized from the samples (Table 1b). The first variety is relatively SiO_2-rich and Al_2O_3-poor. This type is also relatively low in alkalis especially K_2O. The second type is relatively poor in SiO_2, but has high contents of Al_2O_3 and Na (Table 1b).

6.1.3 Feldspar

Feldspar compositions are also variable within the samples. The mineral is present as two-feldspar components, comprising albite and orthoclase in some samples (PN 32A, Table 1c), with representative compositions of 21-32 mol% Ab and 67-78 mol% Or, or as single

Sample:	PN 32A		PN 56	
Analyses no:	1	2	3	4
(a) Calcite				
FeO	0.23	0.26	0.21	0.28
MnO	0.39	0.3	0.45	4.58
MgO	0.02	0.06	0.03	0.01
CaO	53.82	53.3	53.3	51.96
SrO	1.28	1.5	1.33	1.39
Total	55.74	55.42	55.32	58.22
Mg#	15.5	28.3	20.9	4.8

Petrological and Geochemical Characteristics of Mafic Granulites Associated with Alkaline Rocks
in the Pan-African Dahomeyide Suture Zone, Southeastern Ghana

43

Sample:	PN 32A		PN 56		
Analyses no:	1	2	1	2	3
(b) Nepheline					
SiO_2	67.9	65.33	42.69	42.17	42.31
Al_2O_3	20.77	20.37	34.94	34.75	35.17
FeO	0	0.02	0.15	0.04	0.18
CaO	0.54	0.41	0.46	0.69	0.61
Na_2O	11.33	8.5	15.7	15.5	16.06
K_2O	0.08	3.26	6.76	6.51	6.41
Total	100.62	97.89	100.7	99.66	100.74
Si	11.8	11.782	8.165	8.143	8.095
Al	4.255	4.329	7.878	7.908	7.93
Fe	0	0.003	0.024	0.007	0.028
Ca	0.101	0.079	0.094	0.142	0.125
Na	3.819	2.973	5.821	5.801	5.958
K	0.017	0.749	1.65	1.603	1.564
Total	19.992	19.915	23.632	23.604	23.7

Sample:	PN 32A		PN 56	
Analyses no:	1	2	1	2
(c) Feldspar				
SiO_2	67.07	68.05	62.16	61.29
Al_2O3	20.92	20.23	20.38	20.42
FeO	0	0.21	0	0.16
CaO	0.61	0.15	0.1	0.05
Na_2O	11.35	11.79	3.24	2.18
K_2O	0.12	0.07	10.43	12.12
BaO	0	0	3.51	3.87
Total	100.07	100.5	99.82	100.09
Si	2.934	2.964	2.897	2.88
Al	1.079	1.039	1.119	1.131
Fe	0	0.008	0	0.006
Ca	0.029	0.007	0.005	0.003
Na	0.963	0.995	0.293	0.199
K	0.006	0.004	0.62	0.727
Ba	0	0	0.064	0.071
Total	5.011	5.017	4.998	5.017
Mol% An	2.9058	0.6958	0.5447	0.3229
Mol%Ab	96.493	98.9066	31.9172	21.4209
Mol% Or	0.6012	0.3976	67.5381	78.2562

Total Fe as FeO

Table 1. Representative compositions of calcite, nepheline and feldspar in the mafic
granulites from the metasomatic zone.

feldspar comprising almost pure albite with composition of 96-99 mol% Ab (PN 56, Table 1c). A notable feature in the mafic granulites from this study is that calcite, nepheline and feldspars are similar in their compositions to those from the alkaline rocks, with nepheline and feldspars showing similar variability as in the alkaline rocks (Nude et. al., 2009). These comparable features suggest mineralogical influence of the alkaline rocks on the mafic granulites.

7. Geochemistry

7.1 Analytical methods

Whole rock samples were analyzed from representative samples for 10 major elements (SiO_2, TiO_2, Al_2O_3, total Fe as Fe_2O_3*, MnO, MgO, CaO, Na_2O, K_2O, P_2O_5) and 12 trace elements (Nb, Zr, Y, Sr, Rb, Zn, Cu, Ni, Cr, Sc, V, Ba) at Utah State University, and the analytical techniques have been described by Nude et al. (2009). The analysis was carried on Philips 2400 X-ray fluorescence spectrometer using pressed powders for both major and trace elements, with selected U.S.G.S. and international standards prepared identically to the samples. Accepted concentrations were taken from the compilation of Potts et al. (1992). Matrix corrections were carried out within the Philips SuperX software package, which uses the fundamental parameters approach (Rousseau, 1989) to calculate theoretical alpha coefficients for the range of standards. Replicate analyses of selected standards as unknowns suggest percent relative errors $\approx 1\%$ for silica, $\approx 2\text{-}4\%$ for less abundant major elements, and $\approx 1\text{-}6\%$ for trace elements.

The concentrations of rare earth elements (REE) and other trace elements in whole rock samples were determined using Perkin–Elmer 6000 Inductively Coupled Plasma Mass Spectroscopy (ICP-MS) at Centenary College, Shreveport, Louisiana, with acid digestion techniques. Standard reference samples were used in the quantitative analyses of the elements. Table 2 shows the major and trace elements concentrations in the representative samples.

7.2 Major elements

The representative samples of the mafic granulites have SiO_2 contents in the range of 35.0 and 52.0 wt% while CaO contents are from 8.0 to 24.0 wt%. Al_2O_3 contents range from 12.9 to 17.2 wt%; Fe_2O_3 total values range from 7.9 to 10.5 wt% whereas TiO_2 and P_2O_5 are from 1.2 to 1.8 and 0.5 to 0.7 wt% respectively. The total alkalis ($Na_2O + K_2O$) contents are relatively high, with values ranging from 9.7 to 14.1 wt%. Figure 3 are Harker plots in which selected major elements concentrations and total alkalis compositions in the metasomatic mafic granulites are compared to that of the alkaline rocks. The data for the alkaline rocks are from Nude et al. (2009). Apart from K_2O the other major elements from the mafic granulites display linear trends with those from the alkaline rocks. The deviation of K_2O from this trend is not surprising because it is much more mobile and susceptible to alteration. From the present data the linear trends suggest mechanical mixing of the rocks rather than fractional crystallization which can also show linear trend.

Attoh and Morgan (2004) carried out geochemical investigations of the mafic granulites which they sampled from nearby the areas where the present study was carried out, but outside the metasomatic zone, specifically to the east and south of the zone. The following major element ranges (wt %) were reported by these authors: SiO_2 = 42.4-52.0, TiO_2 = 0.9-3.4, Al_2O_3 = 8.6-18.9, Fe_2O_3 total = 6.3- 6.8, MgO = 4.4-11.6, CaO = 7.7-11.1, Na_2O = 1.52- 4.36 + and K_2O = 0.01– 0.57. Their major element results appear similar to those obtained in the present study; exceptions are $Fe_2O_{3total,}$ CaO, the alkalis, Na_2O and K_2O, which are relatively

Petrological and Geochemical Characteristics of Mafic Granulites Associated with Alkaline Rocks
in the Pan-African Dahomeyide Suture Zone, Southeastern Ghana

45

	PN-32A	PN-36	PN-39	PN-42	PN-46	PN-55	PN-56	PN-61	PN-63
SiO_2	35.98	50.03	49.2	38.2	39.33	43.7	42.46	38.4	50.48
TiO_2	1.47	1.37	1.32	1.2	1.31	1.88	1.83	1.32	1.22
Al_2O_3	12.94	17.07	17.11	14.08	14.93	17.23	18.53	14.34	17.21
Fe_2O_3	10.86	8.2	9.41	9.54	9.42	12.34	10.52	9.93	7.92
MnO	0.30	0.28	0.35	0.29	0.29	0.36	0.36	0.31	0.27
MgO	3.05	1.01	1.52	3.71	2.43	1.82	1.48	2.49	0.96
CaO	23.85	8.48	8.01	19.59	17.8	10.25	9.55	19.38	8.36
Na_2O	4.32	8.53	8.95	6.62	7.49	8.24	10.13	6.94	8.64
K_2O	5.25	3.89	2.95	5.14	5.32	2.92	3.97	5.22	3.82
P_2O_5	0.77	0.55	0.64	0.66	0.64	0.58	0.50	0.63	0.56
Total	98.79	99.41	99.46	99.03	98.96	99.32	99.33	98.96	99.44
Mg#	35.8	19.7	24.2	43.5	33.8	22.6	21.8	33.2	19.3
ppm									
Nb	90	256	261	124	146	157	216	144	245
Zr	339	398	365	295	324	266	257	267	352
Y	37	34	36	28	32	46	40	30	36
Sr	3815	2045	1574	3562	3366	1585	1007	3433	1996
Rb	139	82	53	157	139	77	117	137	83
Sc	38	17	16	32	24	16	13	31	14
V	158	60	115	162	191	155	126	199	53
Cr	57	31	21	46	27	33	1	27	44
Ni	12	3	1	23	6	1	1	7	2
Cu	23	4	5	8	17	4	1	16	6
Zn	96	105	111	101	86	87	77	92	106
Ba	3783	1623	1786	2769	2899	2812	2789	2924	1647
La	141.26	124.45	112.01	140.33	114.2	153.2	138.07	146.92	149.81
Ce	258.3	256.77	230.63	222.26	203.32	290.48	275.09	233.75	280.12
Pr	26.46	29.49	25.35	24.28	18.79	31.57	25.17	22.33	34.11
Nd	84.44	99.03	82.34	76.93	59.1	98.99	75.28	68.09	113.59
Eu	3.93	5.03	3.97	3.41	2.84	4.07	2.71	3.32	5.7
Sm	13.33	15.63	12.81	11.59	9.11	13.78	9.63	10.08	17.44
Gd	10.56	11.46	9.53	8.46	6.83	10.01	7	7.42	13.4
Tb	1.26	1.48	1.23	1	0.82	1.29	0.82	0.92	1.72
Dy	6.65	7.35	6.36	4.86	4.15	6.96	4.23	4.42	8.36
Ho	1.25	1.32	1.2	0.89	0.73	1.35	0.87	0.86	1.5
Er	3.2	3.29	2.95	2.25	1.72	3.71	2.37	2.13	3.84
Tm	0.43	0.43	0.41	0.3	0.25	0.53	0.32	0.3	0.54
Yb	2.57	2.49	2.48	1.77	1.4	3.4	1.88	1.74	3.06
Lu	0.38	0.39	0.41	0.26	0.21	0.51	0.29	0.24	0.47
Hf	3.21	3.2	3.37	1.82	1.89	4.25	4.34	1.81	3.32
Ta	4.48	21.4	16.86	6.98	10.44	12.54	18.16	8.64	20.11
Pb	10.43	0.45	0.59	0.35	6.13	-0.76	-1.35	7.08	0.24
Th	4.17	10.99	6.89	4.88	3.77	13.45	12.11	3.62	12.73
U	1.75	4.45	1.55	1.38	2.07	2.13	2.45	1.54	4.2
Zr/Hf	105.6	124.4	108.3	162.19	171.4	62.6	59.2	147.5	106.0
Nb/Ta	20.1	11.0	15.5	17.8	13.0	12.5	11.9	16.7	12.2

Table 2. Major and trace element concentrations in the mafic granulites from the
metasomatic zone.

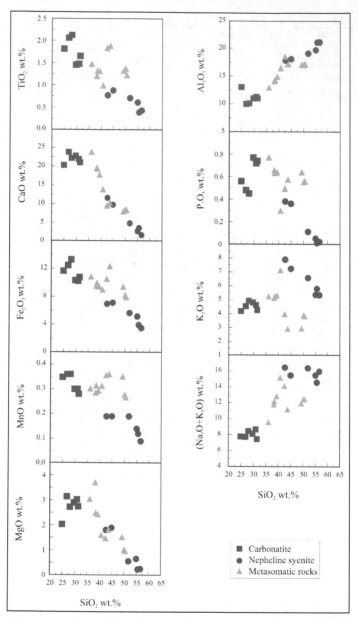

Fig. 3. Harker plots comparing selected major element concentrations in the metasomatic mafic granulite rocks with the alkaline rocks (carbonatite and nepheline syenite). Data for the alkaline rocks are from Nude et al. (2009).

enriched in the rocks from the metasomatic zone compared to those obtained by Attoh and Morgan (2004). For example K_2O contents in the metasomatic rocks are several folds enriched (concentrations range from 2.95 to 5.25 wt %, Table 2) compared to the

Petrological and Geochemical Characteristics of Mafic Granulites Associated with Alkaline Rocks
in the Pan-African Dahomeyide Suture Zone, Southeastern Ghana

47

concentrations in the non-metasomatic varieties (0.01 – 0.57 wt %) determined by Attoh and Morgan (2004). Na_2O also shows similar enrichment in the metasomatic rocks (4.32-10.13 wt %) compared to the non-metasomatic varieties (1.52- 4.36 wt %). The overall major element contents show that the mafic granulites in the metasomatic zone are particularly alkaline, presumably from the addition of Na- and K-rich fluids from the adjacent alkaline rocks. The rocks are also evolved and contain variable amounts of CaO.

7.3 Trace element contents and variations

The trace element contents of the metasomatic zone mafic granulites show high absolute values of Sr (1574-3815 ppm), Ba (1623-3783 ppm), Nb (90-256 ppm). The rocks also have Nb/Ta values ranging from 11-20 and very high Zr/Hf values of 59-171. The Nb/Ta values from the analysed samples compares with chondritic values of 17.6, but the Zr/Hf values are far higher than chondritic values of approximately 36 determined in most reservoirs of the silicate earth (Weyer et al., 2003; Potter, 1996). In Figure 4, Sr and Ba concentrations in the metasomatic mafic granulites are compared with those in the alkaline rocks determined by Nude et al. (2009). From the figure, Sr displays linear trend as is Ba, although 2 samples of the nepheline syenite and a sample of the metasomatic mafic granulites show anomalously high Ba values and deviate from the linear trend. The linear trend has also been shown in the selected major elements in Figure 3, confirming the possible mechanical mixing of those rocks.

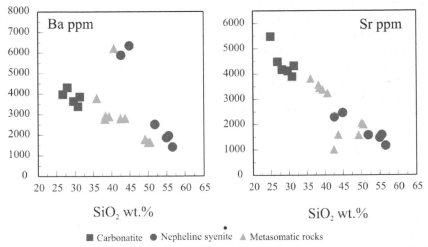

Fig. 4. Ba and Sr concentrations in the metasomatic mafic granulite rocks compared with the alkaline rocks (carbonatite and nepheline syenite). Data for the alkaline rocks are from Nude et al. (2009).

Figure 5 is a primitive mantle-normalized incompatible elements plot of the metasomatic zone mafic granulites compared with alkaline rocks. The similarities of the alkaline rocks to the analyzed samples are shown especially in the relative depletions of the HREE, Rb Hf, Ti, Y and enrichment of K, Eu and Sm relative to Primitive Mantle. The analyzed rocks also differ from the alkaline rocks in elevated U, Ta, Nb, and Zr. Another difference is the prominent troughs at Th, U and Hf, Zr shown in the alkaline rocks; a feature not shown in the mafic granulites. Of particular interest is the relative fractionation of Zr from Hf in the analyzed samples.

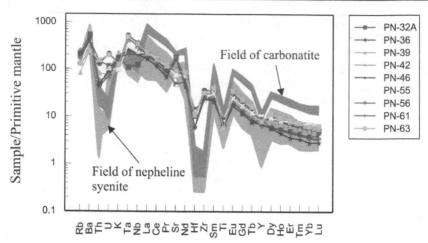

Fig. 5. Primitive mantle-normalized concentrations in the mafic granulites from the metasomatic zone compared with the alkaline rocks (nepheline syenite and carbonatite). Data of the alkaline rocks are from Nude et al. (2009). Normalizing values are from McDonough et al. (1991).

In the REE plot (Fig. 6) the rocks show LREE enrichment, slight positive Eu anomaly and spread at the HREE end. However, the slight positive Eu anomaly is not shown in the carbonatite samples. Overall, the REE patterns of the metasomatic mafic granulites are similar to those of the alkaline rocks, particularly the nepheline syenite. Compared to the suture zone mafic granulites analyzed by Attoh and Morgan (2004) the trace element patterns shown by the rocks from the present study differ markedly. First is the LREE fractionation and steep REE pattern, second is the slight positive Eu anomaly and, third is the overall incompatible trace elements enrichments in the metasomatic granulites.

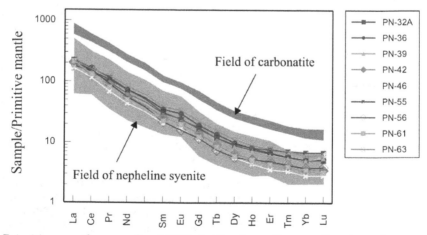

Fig. 6. Primitive mantle-normalized REE plot for the mafic granulites from the suture zone compared with the alkaline rocks (nepheline syenite and carbonatite). Data of the alkaline rocks are from Nude et al. (2009). Normalizing values are from McDonough et al. (1991).

Petrological and Geochemical Characteristics of Mafic Granulites Associated with Alkaline Rocks
in the Pan-African Dahomeyide Suture Zone, Southeastern Ghana

49

8. Discussion

The new data presented in the present study on the modal contents, mineral chemistry and geochemical compositions of the mafic granulite gneiss at the contact zone between the alkaline rocks and the Shai Hills gneiss show that the rocks possess unique compositions which are suggestive of metasomatic transformation of the mafic granulites and hence the modification of the petrography and geochemistry of the original rocks. For example the ubiquitous presence of modal nepheline, feldspar and to a lesser quantity calcite in the mafic granulites is reflective of the mineralogy of the alkaline rocks. Additionally, the common mineral phases such as calcite, nepheline and feldspars in both rock suites have similar mineral chemistry. Partly, the Ca- Na- and K-rich fluids which formed these phases most likely emanated from the alkaline rocks. This evidence is supported by the comparable high CaO, Na_2O and K_2O contents in the mafic granulites under study to the alkaline rocks nearby, and suggests carbonate-alkali fluid interaction.

As shown from the present results the metasomatic mafic granulites are also enriched in Sr, Ba, Nb and show strong REE fractionation; these features are characteristic of carbonatitic melts. Most incompatible trace element concentration and patterns and the REE plots on the mantle-normalized diagrams also show similarities with the alkaline rocks. The observed very high Zr/Hf values in the rocks from the present study which is indicative of Hf fractionation is often associated with carbonate metasomatism (Dupuy et al., 1992) as the element pairs are expected to behave congruently in both fluids and melts (Jochum et al., 1986). Taken together, the major element compositions, the trace element contents and patterns constitute strong evidences to suggest the influence of the alkaline rocks on the overall modal compositions and bulk rock chemistry of the mafic granulites. The linear trends shown in the Harker plots and the spread at the HREE end of the mantle-normalized plots are indicative of mechanical mixing of the rocks, although this requires further evidence to confirm.

However, the metasomatic mafic granulites from the present study preserve some textural, modal and geochemical features which are similar to the Shai Hill gneisses and which also make them different from the alkaline rocks in the area. These features are their coarse texture and presence of garnet, and pyriboles. On geochemistry the mafic granulites again differ in their relative enrichment of Th and U, and depletion of Hf relative to Zr in the spider plots. Importantly the slight positive Eu anomaly in the mantled-normalized plots is absent in the carbonatite. These features provide compelling evidence to suggest that the mafic granulites are alkaline facies of the Shai hills rocks; the trace element budget are likely to have resulted from alkaline fluid interactions.

Available age data show that the emplacement of the alkaline rocks postdates the formation of the mafic granulites (Attoh et al., 2007), so the alkaline rocks are strong candidates for the source of the trace elements and particularly LREE enrichment in the mafic granulites. Although the mechanism of the interaction of the alkaline fluids is not clear, it is possible that it could have resulted from deformation associated with the emplacement of the alkaline rocks through percolation of alkaline fluids and/or mechanical mixing of the rock suites. This hypothesis requires further testing using isotopic data. However, carbonatitic melts have been shown from experiments to be of low viscosity, and capable of separating from their source at low degree melt fractions, and can percolate wall rocks by low angle dihedral flow (Hunter and Mackenzie, 1989; Hammouda and Laporte, 2000). Thus the alkaline fluids are likely to have emanated from the carbonatite and the nepheline syenite into the mafic granulites.

8.1 Tectonic and petrological implications

The new geochemical data from this study also provide further constraints on the evolution of the alkaline rocks and associated mafic granulites exposed along the suture zone of the Pan African Dahomeyide orogen in West Africa. Evidence for the Pan African suture and high pressure metamorphism have been provided in the literature by several workers including Berger et al., (2011), Agbossoumonde, et al. (2001, 2004), Attoh (1998) and Caby, (1987). These data, together with geochemical data on the Shai Hills granulites (Attoh & Morgan, 2004) infer a lower crust chemical composition for the rocks. The latter authors argued that the Dahomeyide mafic granulites preserve chemical imprints of basaltic rocks with trace element compositions similar to those of lower continental crust. The postulation is that the suture zone mafic granulites represent the roots of Pan African volcanic arc that formed from subduction to great depths, followed by HP granulite facies metamorphism accompanied by partial melting and later thrusting along ductile shear zones to produce crystalline nappes along the margin of the West African craton (Attoh, 1998).

The new geochemical data from this study shows a slight positive Eu anomaly in the mantle-normalized trace element patterns in the metasomatic rocks along the suture zone. This feature is interesting because positive Eu anomaly in the Shai Hills rocks has not been previously reported. But evidence of positive Eu anomaly in mafic granulite terrains that formed from basaltic lower crust has been shown by Rudnick (1992). It therefore appears that the mafic granulites from this study preserve a geochemical characteristic that may be of lower crustal affinity, a feature consistent with the findings of Attoh and Morgan (2004). So its tectonic association with the alkaline rocks along the sole thrust of the suture could be partly responsible for the unique alkalic and trace element compositions, as from available age data the alkaline rocks formed possibly after peak granulite metamorphism related to the Pan African orogeny (Attoh et al., 2007).

9. Conclusions

We have provided petrological and geochemical data on the mafic granulites exposed at the contact zone with alkaline rocks of the Kpong complex in the Pan African Dahomeyide suture zone southeastern Ghana. The mafic granulites have been found to have distinct modal and geochemical compositions which are in many ways similar to the alkaline rocks and different from the other mafic granulites outside the contact zone. Some of these are the presence of nepheline-rich veinlets and calcite along cleavage cracks in the mafic granulites. Together with other features such as compositions of modal phases, namely, calcite, nepheline, and feldspar, enrichment of alkalis, Ba, Sr, Nb and LREE, very high Zr/Hf values and overall steep REE patterns observed in the mafic granulites provide compelling evidences which suggest interaction of the alkaline fluids with the mafic granulites at the contact zone.

From the present data the mafic granulites from this study which seem to have a precursor texture and mineralogy identical to the Shai Hill mafic granulite gneisses (Attoh, 1998, Attoh and Morgan 2004), and described by Holms (1974) and Kesse (1985) as mafic nepheline gneiss, is an alkaline facies of the Shai Hills gneisses. The unique composition resulted from alkali fluid interaction from the carbonatitic and nepheline syenite along the tectonized zone. The degree and style of the alkali matasomatism may have varied because of the variable compositions of some the common mineral phases.

Petrological and Geochemical Characteristics of Mafic Granulites Associated with Alkaline Rocks
in the Pan-African Dahomeyide Suture Zone, Southeastern Ghana

51

10. Acknowledgements

This collaboration also forms part of PhD research by PMN. Utah State University provided grants for sample analysis; University of Ghana supported field work for PMN whilst the International Student Exchange Programme provided travel grants. This work is dedicated to the memory of Kodjopa Attoh who passed on when this manuscript was being prepared. His untiring efforts in understanding the Dahomeyides of southeastern Ghana for the past thirty years or so provided the motivation for this work. Reviews by M. M. Ghazal and an anonymous journal reviewer greatly improved the manuscript and are very much appreciated.

11. References

Affaton, P. Kröner, A. & Seddoh, K.F., 2000. Pan-African granulite formation in the Kabye massif of northern Togo (West Africa): Pb-Pb zircon ages. *International journal of Earth Science*, 88, 778-790.

Affaton, P., Rahaman, M.A., Trompette, R., & Sougy, J., 1991. The Dahomeyide orogen: Tectonothermal evolution and relationship with the Volta basin. In: *The West African orogens and circum-Atlantic correlatives*, Dallmeyer, R.D, Lecorche, J.P. (Eds.), 95-111 Springer, New York.

Agbossoumonde , Y., Guillot , S. & Ménot , R. P. 2004. Pan-African subduction collision event evidenced by high-P corona in metanorites from Agou massif (southern Togo). *Precambrian Research*, 135, 1–25.

Agbossoumonde, Y., Ménot, R.-P. & Guillot, S. 2001. Metamorphic evolution of Neoproterozoic eclogite from south Togo (West Africa). *Journal of African Earth Sciences*, 33, 227–244.

Agbossoumonde, Y., Ménot, Pacquette J.L., Guillot, S, Yessoufou S., & Perrache C.,2007. Petrological and geochronological constraints on the origin of Palimé-Amlamé granitoids (South Togo, West Africa): A segment of the West African craton Palaeproterozoic margin reactivated during Pan-African collision. *Gondwana Research*, 12, 4750-488.

Agyei, E.K., van Landewijk, J.E.J.M., Armstrong, R.L., Harakal, J.E., & Scott, K.L., 1987. Rb–Sr and K-Ar geochronometry of south-eastern Ghana. *Journal of African Earth Sciences*, 6, 153–161.

Attoh, K., 1998. High-pressure granulite facies metamorphism in the Pan-African Dahomeyide Orogen, West Africa. *Journal of Geology*, 106, 236–246.

Attoh, K., & Morgan, J., 2004. Geochemistry of high-pressure granulites from the Pan-African Dahomeyide orogen, West Africa: constraints on the origin and composition of lower crust. *Journal of African Earth Sciences*, 39, 201-208.

Attoh, K., & Nude, P.M. 2008. Tectonic significance of carbonatite and ultrahigh-pressure rocks in the Pan-African Dahomeyide suture zone, southeastern Ghana. In: *The boundaries of the West African craton*, Ennih, N., Liégeois, J. P (eds.), Geological Society of London Special. Publication, 297, 217-231.

Attoh, K., & Smith, M. D. 2005. Nd and Hf isotopic compositions of Pan-African high-pressure mafic granulites. EOS Transactions, American Geophysical Union 86 (18) Joint Assembly Supplement V13B-02.

Attoh, K., Corfu, F., & Nude, P.M., 2007. U–Pb zircon age of deformed carbonatite and alkaline rocks in the Pan-African Dahomeyide suture zone, West Africa. *Precambrian Research*, 155, 251–260.

Attoh, K., Dallmeyer, R.D., & Affaton, P., 1997. Chronology of nappe assembly in the Pan-African Dahomeyide orogen, West Africa: evidence from 40Ar/39Ar mineral ages. *Precambrian Research*, 82, 135–171.

Attoh, K., Hawkins, D., Bowring, S., & Allen, B., 1991. U-Pb zircon ages of gneisses from the Panafrican Dahomeyide Orogen, West Africa. *EOS Transactions*, American Geophysical Union, 72, 229.

Bell, K., & Tilton, G.R., 2001. Nd, Pb and Sr isotopic compositions of east African Carbonatites: evidence for mantle mixing and plume inhomogeneity. *Journal of Petrology*, 42, 1927–1945.

Berger, J., Caby, R., Liégeois, J-P., Mercier, C. J-C., & Demaiffe, D., 2011. Deep inside a neoproterozoic intra-oceanic arc: growth, differentiation and exhumation of the Amalaoulaou complex (Gourma, Mali). *Contributions to Mineralogy and Petrology*. DOI: 10.1007/s00410-011-0624-5.

Bernard-Grifiths, J., Peucat, J. J., & Menot, R. P., 1991. Isotopic (Rb-Sr, U-Pb, and Sm-Nd) and trace element geochemistry of eclogites from the Pan-African belt: a case study of REE fractionation during high grade metamorphism. *Lithos*, 27, 43-57.

Caby, R., 1987. The Pan-African belt of West Africa from the Sahara to the Gulf of Guinea. In: *Anatomy of Mountain Ranges*, Schaer, J.P., Rodgers, J. (Eds.), 129-170. Princeton University Press,.

Castaing, C., Triboulet, C., Feybesse, J-L., & Chevrement, P., 1993. Tectonometamorphic evolution of Ghana, Togo, and Benin in the light of the Pan-African/Brasiliano orogeny. *Tectonophysics*, 218, 323-342.

Chakhmouradian, A. R., Mumin, A. H., Deméy, A., & Elliott, B., 2007. Postorogenic carbonatites at Eden Lake, Trans-Hudson Orogen (northern Manitoba, Canada): geological setting, mineralogy and geochemistry *Lithos*. doi:10.1016/j.lithos.2007.11.004.

Cordani, U.G., D'Agrella-Filho, M.S., Brito-Neves, B. B., & Trindale, I.F., 2003. Tearing up Rodinia: the Neoproterozoic paleogeography of South American cratonic fragments. *Terra Nova*, 15, 350-359.

Dawson, J. B., Penkerton H., Norton G. E., & Pyle D. M., 1990. Physicochemical properties of alkali carbonatite lavas: Data from the 1988 eruption of Oldoinyo Lengai, Tanzania, *Geology*, 18, 260-263.

Dupuy, C., Liotard, J. M., & Dostal, J. 1992. Zr/Hf fractionation in intraplate basaltic rocks: carbonate metsaomatism in the mantle source. Geochimica et Cosmochimica Acta 56, 2417-2423.

Hammouda, T., Laporte, D., 2000. Ultrafast mantle impregnation by carbonatite melts, Geology 28, 283–285.

Hirdes, W. & Davis, D. W. 2002. U–Pb zircon and rutile metamorphic ages of the Dahomeyan garnet– hornblende gneiss in southeastern Ghana, West Africa. *Journal of African Earth Sciences*, 35, 445–449.

Hoffman, P.F., 1991. Did the breakout of Laurentia turn Gondwana inside-out? *Science*, 252, 1409–1412.

Petrological and Geochemical Characteristics of Mafic Granulites Associated with Alkaline Rocks
in the Pan-African Dahomeyide Suture Zone, Southeastern Ghana

53

Holm, F.R., 1974. Petrology of alkalic gneiss in the Dahomeyan of Ghana. *Geological Society of America Bulletin*, 85, 1441–1448.

Hornig-Kjarsgaard, I., 1998. Rare earth elements in sovitic carbonatites and their mineral phases. *Journal of Petrology*, 39, 2105–2121.

Hunter, R.H., MacKenzie, D., 1989. The equilibrium geometry of carbonate melts in rocks of mantle composition. *Earth and Planetary Science Letters*, 92, 347–356.

Jochum, K. P., Seufert, H. M., Spettel, B., & Palme, H., 1986. The solar system abundances of Nb, Ta, and Y, and the relative abundances of refractory lithohpile elements in differentiated planetary bodies. Geochemica et *Cosmochemica Acta* 50, 1173–1183.

Kesse, G. O., 1985. *The mineral and rock resources of Ghana*. A.A. Balkema Publishers,ISBN 9061915899, Rotherdam.

McDonough, W. F., Sun, S., Ringwood, A. E., Jagoutz, E., & Hofmann, A. W. 1991, K, Rb, ans Cs in the earth and moon and the evolution of the earth's mantle, *Geochimica et Cosmochimca Acta*, Ross Taylor Symposium volume.

Nelson, D. R., Chivas, A R., Chapell, B. W., & McCulloch, M T., 1988. Geochemical and isotopic systematics in carbonatites and implications for the evolution of ocean island sources. *Geochemica et Cosmochemica Acta*, 52, 1–17.

Nude P.M. ,Shervais J., Attoh K., Vetter S. K., & Barton C., 2009 Petrology and geochemistry of nepheline syenite and related carbonate-rich rocks in the Pan- African Dahomeyide orogen, southeastern Ghana, West Africa. *Journal of African Earth Sciences*, 55, 147-157.

Nude, P. M., Corfu, F. & Attoh, K. 2006. U–Pb zircon ages of deformed carbonatite and alkaline rocks in the Pan-African Dahomeyide suture zone, West Africa. *EOS Transactions*, American Geophysical Union, 87, Fall Meeting Supplement, V31B-0585.

Potter, L. E., 1996. Chemical variation along strike in feldspathoidal rocks of the eastern alkali belt, trans-pecos magamatic province, Texas and New Mexico. In: *Alkaline rocks: petrology and mineralogy*, Mitchel, R. H., Eby, G. N., & Martin, R. F. (Eds)., *Canadian mineralogist*, vol. 34 241-263.

Potts, P. J., Tindle, A. G., & Webb, P C., 1992. Geochemical Reference Material Composition. CRC Press, Boca Raton, FL. 313pp.

Rousseau, R. M., 1989. Concepts of influence coefficients in XRF analysis and calibration. In: Ahmedali, S.T. (Ed.), X-Ray Fluorescence Analysis in the Geological Sciences: Advances in Methodology. *Geological Association of Canada GAC-MAC Short Course*, vol. 7, pp. 141–220.

Rudnick, R. L. 1992. Restites, Eu anomalies, and the lower continental crust, *Geochimica et Cosmochimica Acta*, 56, 963-970

Rudnick, R. L., McDonough, W. F., & Chappell, B. W. 1993. Carbonatite metasomatism in the northern Tanzanian mantle: petrographic and geochemical characteristics. *Earth and planetaryscience letters*, 114, 463-475.

Sylvain J. P., Aregba A., Collart J., & Godonou K. S.. 1986. Notice explicative de la carte géologique du Togo 1//500,000. *Direction Generale des Mines de la Géollogie et du Bureau National de Recherches Minières*. Memoire 6.

Thompson, R. N., Smith, P. M., Gibson, S. A., Mattey, D. P., & Dickin, A. P., 2002. Ankerite carbonatite from Swartbooisdrif, Namibia: the first evidence for magmatic ferrocarbonatite. *Contributions to Mineralogy and Petrology*, 143, 377–395.

Tohver, E., D'Agrella-Filho, M. S. & Trindale, R. I. F., 2006. Paleomagnetic record of Africa and South America for 1200-500 Ma interval, and evaluation of Rodinia and Gondwana Assemblies. *Precambrian Research*, 147, 193-222.

Wallace, M. E., & Green, D. H., 1988. An experimental determination of primary carbonatite magma composition. *Nature*, 335, 343–346.

Weyer, S., Munker, C., & Mezger, K., 2003. Nb/Ta, Zr/Hf and REE in the depleted mantle: implications for the differentiation history of the crust-mantle system. *Earth and Planetary Science Letters*, 205, 209–324.

Winter, J. D., 2001: An Introduction to Igneous and metamorphic petrology, Prentice Hall Inc, 697pp.

Woolley, A. R.., & Kemp, D. R. C., 1989. Carbonatites: nomenclature, average chemical compositions, and element distribution, In: *Carbonatites – Genesis and Evolution*, Bell, K. (Ed.), Unwin Hyman, London.

Petrogenesis and Tectono-Magmatic Setting of Meso-Cenozoic Magmatism in Azerbaijan Province, Northwestern Iran

Hemayat Jamali[1,2], Abdolmajid Yaghubpur[2], Behzad Mehrabi[2],
Yildirim Dilek[3], Farahnaz Daliran[4] and Ahmad Meshkani[2]

[1]*Geological Survey of Iran, Tehran*
[2]*Tarbiat Moallem University, Tehran*
[3]*Department of Geology, Miami University, Oxford, OH*
[4]*Institute for Applied Geosciences, University of Karlsruhe, Karlsruhe*
[1,2]*Iran*
[3]*USA*
[4]*Germany*

1. Introduction

There are widespread Cenozoic magmatic rocks in the prei-Arabian part of northern Zagros suture zone, which form the continental crust of this part of Alpine – Himalayan orogenic belt (Fig. 1). Although the age of these rocks ranges from Cretaceous to Quaternary, but the main magmatic phases belong to Cretaceous, upper Eocene- Oligocene, upper Miocene-Pliocene and plio-Quaternary. The magmatism occurred in Meso-Cenozoic periods due to convergence of the Arabian and Eurasian plates. The time and space distribution of Meso-Cenozoic magmatism in Azerbaijan between Arabian and Eurasian plates and their tectonic setting, are the major question in geodynamics of the eastern Mediterranean and Eastern Alpine-Himalayan belt.

In this paper, new geochemical data of Cenozoic magmatic rocks of Arasbaran region (analyzed by XRF method in Miami University) are used for interpreting petrogenesis of these rocks and determining their origin, nature of magma and its evolution. The time and space distribution of Late Mesozoic- Cenozoic magmatic rocks of the study area are also compared with the adjacent region in Azerbaijan, Armenia and Eastern Turkey. The authors try to find the relationship between magmatism and mineralization in time and space and to locate the probable occurrence of different types of mineralization associated with various magmatic events in the study area.

2. Regional geology

The broad Tethyan orogen had been evolved during a series of successive collisions between Eurasia and the rifted fragments of Gondwana land (Shengor and Natal'in, 1996). Rifting of ribbon-like continental fragments (i.e. Central Iran, South Armenia and Tauride blocks) from Gondwana occurred in the late Paleozoic–early Mesozoic, and discrete Tethyan ocean

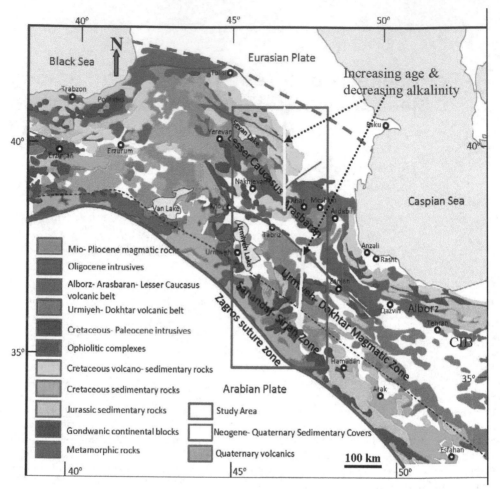

Fig. 1. Geological map of Tethyan belt from Central Iran to Eastern Turkey (modified from Aghanabati, 1993).

basins developed in the wake of these northward-migrating continental blocks (Dilek et al., 2010). The results of the collided segments of Gondwana land and Arabian plate with Eurasian plate were the construction of Zagros thrusted mountain range trending NW- SE in the western part of Iran, and Alborz-Azerbaijan mountain range in the northern to northwestern parts of the country. Four major tectono-magmatic zones; Sanandaj-Sirjan zone (SSZ), Urumieh- Dokhtar magmatic belt (UDMB), Central Iran Block (CIB) and Alborz-Arasbaran- Lesser Caucasus Belt (AALCB) in northwest of Iran are the result of geodynamic evolution of Tethys belt formed between Arabian and Eurasian plates during Early Mesozoic to Late Cenozoic (Fig. 1).

These Tectono-magmatic zones could be divided into smaller magmatic-metallogenic subzones such as Songhor-Baneh subzone, Sanandaj Cretaceous volcanic subzone and Tabriz-Hamadan subzone situated in the tectono-magmatic zone between Tabriz fault and

Zagros thrust fault (Azizi, 2009). Blourian (1994) divided the magmatism of northern Iran into two subzones: Alborz subzone and West Alborz- Azerbaijan subzone. Jamali et al. (2010) recognized three magmatic–metallogenic subzones in the Ahar-Arasbaran–Lesser Caucasus area.

2.1 Sanandaj-Sirjan Zone (SSZ)

Sanandaj-Sirjan metamorphic-magmatic zone trending NW-SE is extending from south of Iran to southeast of Turkey parallel to the main Zagros thrust fault. Many geologists believe that the subduction of Arabian plate under the SSZ had been occurred in the place of main Zagros thrust fault where the ophiolites are situated along the thrust (Takin, 1972; Dewey et al., 1973; Berberian and Kings, 1981; Shengor, 1990; Hesami et al., 2001; Talebian and Jakson, 2004). On the other hand, Alavi (2007) believes that the suture zone between the Arabian plate and Iran lies between the SSZ and UDMB.

The northern part of the SSZ includes upper Paleozoic and Mesozoic metamorphic rocks with Meso-Cenozoic intrusive bodies. The age of these intrusions (Fig. 2) ranges from 170 to 40 Ma (Ghalamghash, 2009; Mehrabi et al., 2009; Ghaderi et al., 2009). These bodies are composed of granite, granodiorite, syenite and diorite that belong to medium potassium calc-alkaline series, and from genetic point of view, they belong to I, A and S type granitoids, and are formed during syn- to post-collision environments in subduction zones (Ghalamghash, 2009; Mazhari et al., 2009). In addition to intrusive events, two volcanic subzones were also recognized in SSZ (Azizi, 2009). Songhor–Baneh subzone (5 to 10 km wide and extending about 200 km) is composed of volcanic–sub-volcanic facies including basalt, gabbro and diorite with tholeiitic to calk-alkaline characteristics. In some places, the rocks are metamorphosed to greenschist and amphibolite facies. The age of these volcanic rocks using K-Ar dating method (Moinvaziri et al., 2008) is Late Eocene–Miocene (42-27 Ma). Azizi et al. (2009) believed that the Baneh-Songhor subzone that formed between the ophiolite suture zone (Campanian–Maastrichtian) and the SSZ belongs to an oceanic arc and it is the result of subducted Neo-Tethys oceanic crust under another fragment of the oceanic crust.

Another volcanic subzone belongs to the Late Cretaceous including mafic to intermediate rocks of calc-alkaline affinity. This subzone with 15-20 km width and 300 km length extends from Saghez to Piranshahr.

Differentiated REE pattern (Fig. 5), low Ti, high Al and negative Nb anomaly are indicative of continental margin subduction zone (Azizi, 2009). Barika barite- and gold-rich massive sulfide deposit is associated with this volcanism (Tajeddin, oral communication).

According to Ghalmghash (2009), the subduction of Arabian plate under the SSZ had been started in the Early Cretaceous, and the first magmatic activities are related to the subduction zone. He believes that the magmatism that occurred after Campanian (80 Ma) is not related to subduction and it is originated from the crust, formed due to the collision between Arabian margin and the SSZ. Mazhari (2009) believes that the Piranshahr intrusive bodies of 41 Ma belong to post collision event and the age of the collision between Arabian plate and the SSZ occurred in Late Cretaceous, while Omrani et al. (2008) and Azizi (2009) think that the collision was much younger and it occurred in the Late Miocene.

Gold mineralization (55.7-38.5 Ma) in metamorphic rocks (schist and amphibolite) of the SSZ (e.g. Muteh deposit and Saghez region) shows intrusion-related characteristics rather than orogenic mineralization. Magmatic phases, which are simultaneous with mineralization, are reported in the SSZ (Moritz et al., 2006; Tajeddin, oral communication).

There is no mark about the porphyry and epithermal mineralization associated with Mesozoic-Paleogene magmatism in the SSZ.

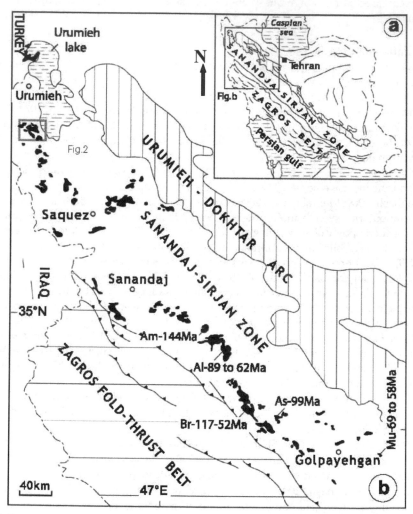

Fig. 2. (a) Tectonic setting of the Zagros orogenic belt in western Iran (b) Distribution of the Urumieh-Gholpaiegan plutonic belt (in black) in the SSZ. The ages of plutonic bodies are given in Ma, (Ghalamghash, 2009).

2.2 Urumieh-Dokhtar magmatic belt(UDMB)
Urumieh-Dokhtar magmatic belt trending NW-SE parallel to the SSZ (50 to 80 km wide) extends from south to northwest of Iran, situated next to Tabriz fault in northeast of this zone. There are some volcanic rocks cropping out in the south of Sarab. Magmatic activities in this belt started in the Eocene and continued until Quaternary (Omrani, 2008; Azizi, 2009). Magmatic activities are divided into two phases: a) volcanic activities in Eocene followed by

intrusive activities in Oligocene, and b) magmatic activities started in the Late Miocene and continued to Quaternary. Data of major and trace elements in UDMB indicate the characteristics features of calc-alkaline magmatism related to continental margin subduction zones. REE pattern of these rocks is also indicative of mantle origin for the Eocene volcanic rocks (Omrani, 2008). On the contrary, the patterns of major and minor elements of the Late Miocene volcanic rocks, with adakite characteristics, are indicative of the break off of subducted slab and post collision process (Omrani, 2008). Omrani (2008) also believes that the Late Miocene collision and breakage of subducted slab in UDMB is simultaneous with the same event in the southeast of Turkey that occurred in 5-10Ma ago (Keskin, 2003).

Eocene volcanic rocks are associated with manto-type copper mineralization (e.g. Sarab and Mianeh regions). Recent studies indicate the probable existence of porphyry type mineralization in the Hashtrud- Mianeh region accompanying Oligocene plutonism that intruded the Eocene volcanic rocks (Jamali, 2011; unpublished data). Miocene magmatism with an adakitic signature and mildly alkaline affinity is accompanied with epithermal and Carlin type gold mineralization in Ghorveh and Takab regions (Mehrabi et al., 1999; Richards, 2006 and 2009).

2.3 Central Iran Block (CIB)

There are scattered exposures of the CIB, overlain by the Cenozoic volcanic rocks in northeast of UDMB and Alborz-Azerbaijan mountain ranges. These rocks, cropping out in the study area, are pre-Triassic sedimentary rocks that are similar to those of the south Armenia and Tauride blocks in Turkey. According to Sosson et al. (2005) and Dilek et al. (2010), Central Iran, Armenia and Tauride blocks were formed the northern margin of the Gondwana land prior to Triassic rifting. These three blocks include Precambrian crystalline basement overlain by Paleozoic–Mesozoic sedimentary sequences (Dilek, 2009; Roland, 2009; Sosson, 2005), separated from Gondwana land in the Late Triassic and joined the southern margin of the Eurasian plate during Late Cretaceous–early Paleocene (Dilek, 2009; Roland, 2009). Paleomagnetic investigations on Middle Triassic igneous rocks in Sumkhit- Gharebagh (Azerbaijan) and south Armenia indicate that the latitude was at about 22° N for the time of their formation (Sosson, 2005). The Paleo-Tethys Ocean started closing as the Neo-Tethys Ocean opened during separation of Central Iran, Armenia and Tauride blocks from the Gondwana land moving northwards. According to Bazhenov et al. (1996), the site of paleo-Tethys ocean enclosure is the present place of Sumkhit-Gharabagh-Black Sea area.

2.4 Alborz-Arasbaran-Lesser Caucasus Belt (AALCB)

The Transcaucasian Massif includes Pan-African orogenic crust intruded by latest Proterozoic to Palaeozoic granitoids, experienced multiple deformation and migmatitization, and the Jurassic to Early Cretaceous plutons representing a magmatic arc (Zakaridze et al., 2007). This arc continued into the Eastern Pontide blocks in the west.

A Cretaceous island arc complex with calc-alkaline to alkaline extrusive rocks, and pyroclastic deposits, flysch units and marl-limestone rocks occurs north of the Sevan-Akera suture zone (Dilek et al., 2010).

Cenozoic magmatic rocks of the AALCB are mainly acidic to basic volcanic rocks and acidic to intermediate intrusive bodies, with calc-alkaline to alkaline affinities trending NW-SE and located in the northwestern Iran and Caucasus mountain ranges (Moayed, 2001; Jamali et al., 2010). Similar to UDMB, magmatic activities in this belt have been started in the Eocene and continued to Quaternary.

Azizi et al. (2009) and Alavi (2007) believe that the scattered pieces of ophiolites around Tabriz fault are the remnants of UDMB back-arc oceanic crust (basin) and its subduction was the cause of the formation of Alborz-Arasbaran magmatic belt. There are scattered exposures of ophiolites and metamorphic rocks from Anzali to Sevan- Akera at northern margin of the AALCB (Fig. 1) (Galoyan et al., 2009; Berberian et al., 1981; Jamali et al., 2010). Based on isotopic data, the formation of Sevan-Akera ophiolites have started in the Middle-Late Jurassic and continued to Early Cretaceous (Galoyan et al., 2009). According to Sosson (2005), based on interlayering of radiolarites and pillow basalts, the age of ophiolites in the northern Armenia is proposed to be Late Jurassic. The exposure sites of these ophiolites could be the site of the Mesozoic ocean that was closed in Late Cretaceous-Early Paleocene causing the AALCB magmatisms. Jurassic to Cretaceous magmatic rocks with alkaline to calc-alkaline affinities in the north Lesser Caucasus indicates arc magmatism characteristics and continued to the west joining the Pontide magmatic belt in northeastern Turkey (Dilek, 2010; Zakaridze, 2007). Intense volcanic activities with acidic to basic composition were started in the AALCB during Eocene. Eocene volcanic rocks with alkaline to calc-alkaline nature (Figs. 3 and 4) indicate continental margin magmatism. Distribution patterns of the

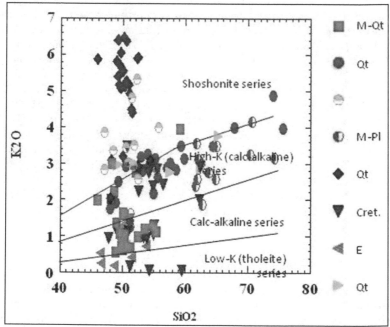

Fig. 3. K_2O (wt%) versus SiO_2 (wt%) diagram for clasification of NW Iran volcanic rocks (Peccarillo and Taylor, 1976). The abbreviations are as follow: M-Qt: Mio-Pliocene to Quaternary(Kheirkhah et al., 2009), Qt: Quaternary (Dilek et al., 2010), E: Eocene (Dilek et al., 2010), M-Pl: Mio-Pliocene (Dilek et al., 2010), Qt: Quaternary (Ahmadzadeh et al., 2010), Cret: Cretaceous (Azizi et al., 2009), E: Eocene (Azizi et al., 2009), Qt: Quaternary (Ahmadzadeh et al., 2010).

REEs (weak or absent Eu negative anomaly and more differentiated pattern of REEs) also are indicative of continental subduction zones (Fig. 5). Most of the samples (except some Cretaceous volcanics in SSZ) plotted in $(Nb/Zr)_n$-Zr - Zn diagram clustered in the collision related and within plate fields (Fig. 6). In Rb - Y+Nb and Nb - Y diagram (Pearce, 1996, 1984) they mainly plotted in the field of post collision granitoids (Fig. 7). In the Hf, Nb, Ta and Rb diagrams of Harris (1986), most of the samples plotted at the border of VAG, WPG and Post-Collision fields (Fig. 8).

Following to the Eocene intense volcanic activities, Oligocene magmatic activities occurred in the form of intrusive bodies, in the AALCB. The composition of Oligocene intrusions is mainly acidic to intermediate with lesser amounts of mafic rocks (Fig. 9). They include granite, granodiorite, quartz monzonite, monzonite, syenite, syeno-diorite and gabbro which are intruded by lamprophyric to dacitic dykes. Based on the geochemical studies, the intrusive rocks are of different origins. Monzodiorites, gabbros and syenites show alkaline and shoshonitic nature and probably were originated from partial melting of lithospheric mantle, while the acidic rocks (granites, granodiorites and quartz monzonites) with high-K, calc-alkaline affinities (Fig. 8) show C type adakitic magma characteristics. They also show high amounts of Ba and Sr and originated from partial melting of lower mafic and potassic crust due to increase of the crust's thickness. Finally, the lamprophyrs are alkaline type and formed from the melting of OIB type metasomatized lithospheric mantle (Aghazadeh, 2010).

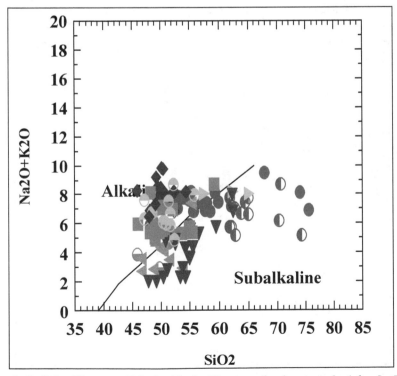

Fig. 4. Total alkali (wt%) versus SiO_2 (wt%) clasification of volcanic rocks (after Le Bas et al., 1986). (symbols as in Fig.3)

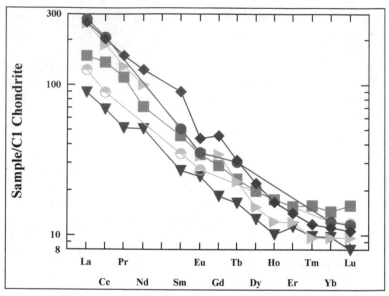

Fig. 5. C1 Chondrite normalized REE patterns for Meso-Cenozoic volcanic rocks in NW Iran. Mean values of REE used for each rock group. (symboles as in Fig. 3).

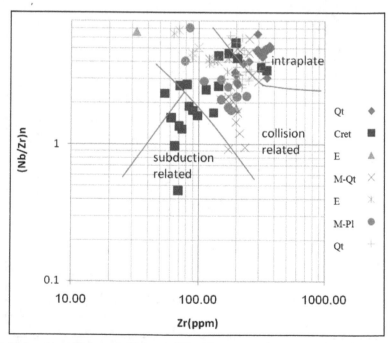

Fig. 6. (Nb/Zr)ₙ versus Zr (ppm) diagram (Thieblemont and Tegyey, 1994) for volcanic rocks. Most of the samples plotted in the collision related field. (Abbreviations as in Fig. 3).

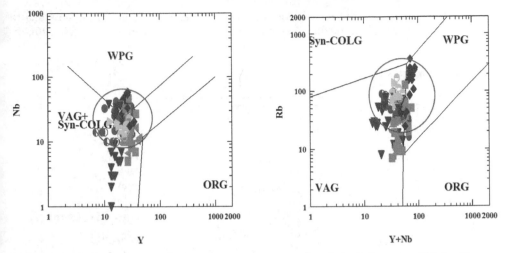

Fig. 7. Tectonic discrimination diagram of Rb versus Yb + Nb and Nb versus Y (after Pearce et al., 1984 and 1996). Most of the samples plotted in the post collision field (blue circles). (symbols as in Fig.3)

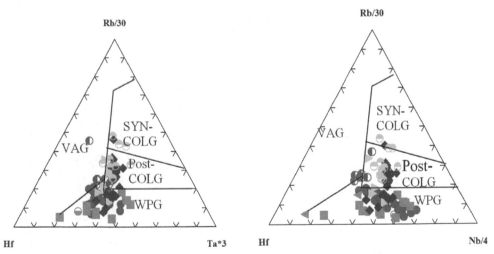

Fig. 8. Triangular diagrams of Harris et al. (1986), for various tectonic environments in collision zones. Most of the samples plotted at the border of VAG, WPG and Post-Collision field, which is characteristics of the post-collision magmatism. (symbols as in Fig. 3)

Field observations indicate that the acidic calc-alkaline rocks are older than intermediate to basic shoshonitic rocks. These observations are confirmed by zircon age dating of intrusive bodies of north and east of Ahar given by Aghazadeh (2010). The results of these dating indicate that the age of adakitic granodiorite is 31.8 Ma., gabbro-monzonite is 28 Ma and syenite and melasyenite is 24-26 Ma. In K_2O-SiO_2 diagram, lamprophyrs and synenites are plotted in shoshonite field and gabbros, monzonites and granites (resulted from their differentiation) belong to the high potassium series. Granites, granodiorites and quartz monzonites are situated in medium to high potassium calc-alkaline field (Fig. 10). In diagram of trace elements of Pearce (1984, 1996) and Thieblemont and Tegyey (1994), most of the samples belong to post collision environments and they have no direct relationships with subduction (Figs. 11 and 12).

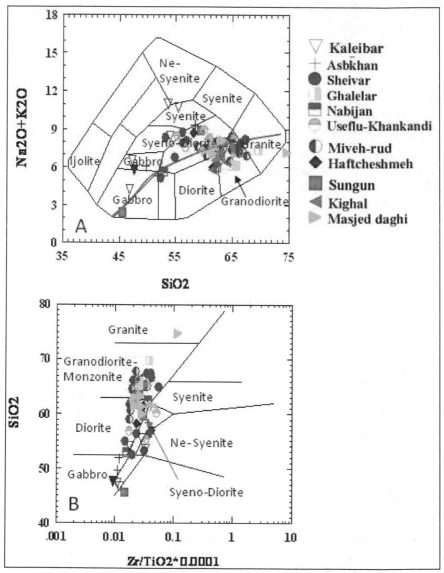

Fig. 9. Geochemical classification of the AALC plutons. (A) Using the total alkali versus SiO_2 (Wilson, 1989); the blue line separates alkaline rocks from sub-alkaline. (B) Zr/TiO_2 versus SiO_2 diagram of Winchester and Floyd (1977).

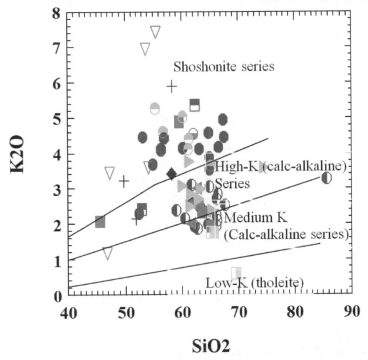

Fig. 10. K_2O (wt%) versus SiO_2 (wt%) diagram for clasification of the Arasbaran Oligicene plutonic rocks (Le Maitre et al.,1989; Rickwood, P.C., (1989). (symboles as in Fig.9).

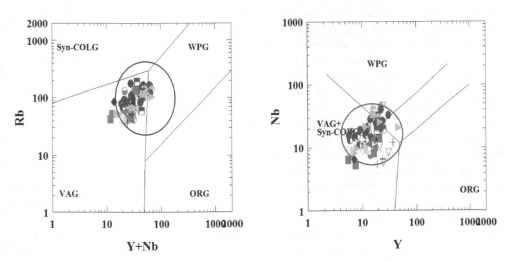

Fig. 11. Tectonic discrimination diagram of Rb versus Yb + Nb and Nb versus Y (after Pearce et al., 1984; Pearce, 1996). Most of the samples plotted in post collision field (blue circle). (symboles as in Fig.9).

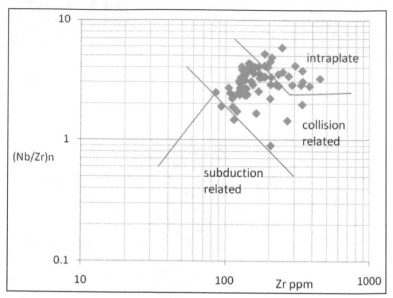

Fig. 12. $(Nb/Zr)_n$ versus Zr (ppm) diagram (Thieblemont and Tegyey, 1994) for the volcanic rocks. Most of the samples plotted in the collision related field.

Oligocene intrusive bodies could be divided into two groups based on their structure, texture and the depth of emplacement: (1) Large intrusive bodies (batholites) with medium to coarse grained texture that formed at 3 to 4 km depth such as Sheivar Dagh, Khankandi, Youseflu and Kaleibar are poorly mineralized, and mineralization associated with them mainly occurs in the contact zone with their host rocks or within the country rocks. These intrusive bodies show high K calc-alkaline to shoshonitic affinity. (2) Sub-volcanic bodies with porphyritic textur and calc-alkaline character formed at 1-2 km depth and are associated with remarkable mineralization such as Sungun, Mivehrud, Masjed Daghi and Haftcheshmeh deposits. Most of the mineralization occurs in quartz monzonite-monzonite. Microscopic investigations of rock samples of quartz monzonite- monzonite indicate that amphibole and biotite phenocrysts are paragentically earlier than other minerals. This indicates that the amount of water in the original magma was more than 3% in the crystallization stage (Whitney, 1975; Burham, 1979; Whitney and Stromer, 1985). The lack of Eu negative anomaly may be assumed to be the result of high oxygen fugacity in the original magma, because the high oxygen fugacity oxidizes Eu^{2+} to Eu^{3+} and so it can not enter into the plagioclase structure (Hezarkhani, 2006). In addition to bearing water and oxidizing state of the primary magma, replacement in the lower depth caused water saturation and fluid separation in primary stages of crystallization of original magma followed by mineralization.

After Oligocene intrusion phase in the Arasbaran area the next magmatic activities occurred in the Late Miocene and continued to Quaternary. The Late Miocene-Pliocene magmatic rocks are mostly acidic in composition (granite–granodiorite) and they are in the form of sub-volcanic bodies and/or volcanic domes (such as Ghalehlar and Bohlul Daghi), that could be observed in the southwestern Arasbaran area. These magmatic rocks have alkaline adakitic nature and they are probably related to post collision processes such as slab break off or delamination (Jahangiry, 2007; Karimzade, 2004). Adakitic magmatism in the Plio-Quaternary was followed by basic alkaline volcanism. At first, the alkali basic magma was sodic in composition changing upward to more potassic in composition, and bears leucite phenocrysts. The Plio-Quaternary volcanic rocks include andesite, basalt-andesite, tephrite-basalt, leucite basalt and tephrite. The amounts of K_2O and MgO and the K_2O/Na_2O ratio in these rocks are high enough to classify them as high or even ultra potassic rocks (Fig. 3). They characterized with significant enrichment in LILE and LREE and depletion in high-field-strength elements. So, the metasomatized lithospheric mantle composed of garnet lherzulite and high fugacity of CO_2/H_2O ratio played a significant role in their genesis (Ahmadzadeh, 2010 and Khezerlu, 2008).

2.5 Discussion and conclusions

Two oceanic island arcs magmatism, one in SSZ (Songhor-Baneh) and the other in the north Lesser Caucasus are distinguished. Except of these two oceanic arcs, all of the magmatic rocks in the SSZ, UDMB and AALCB are related to continental margins. As going toward north and south from Tabriz fault, the magmatism becomes older and its alkalinity decreases.

SSZ magmatic rocks that are mostly intrusive type were primarily related to an active subduction margin and later on about 80 million years ago reveals post–collision magmatic characteristics. There is no unanimity for the time of collision between the Arabian plate and SSZ. Some geologists believe that the collision had been occurred in the Late Cretaceous–Early Paleocene (Moaiyed, 2001; Ghalamghash, 2009; Mazhari, 2009), while some others think of the collision had been occurred during Middle–Late Miocene (Omrani, 2009; Azizi, 2009). However, the rare earth and trace elements diagrams indicate that the major plutonic and volcanic rocks of the UDMB and SSZ belong to the post-collision environment. So, it may be concluded that the time of collision is more likely Late Cretaceous–Early Paleocene, and the following magmatisms are not directly related to subduction. On the other hand they may be related to the post collisional magmatism such as slab break off and delamination.

From Tabriz fault toward north to Lesser Caucasus, a reverse zonation is observed, that means the age of magmatic rocks increases toward north. In the vicinity of Tabriz fault the magmatic rocks are of Miocene–Quaternary age, while in the Arasbaran region and southern part of Lesser Caucasus, the magmatism belongs to Eocene- Oligocene, and from Sevan-Akera suture zone toward north, the magmatic rocks are related to Jurassic-Late Cretaceous events. However, in the study area the older rocks (Late Mesozoic) show magmatic arc characteristics, while the younger rocks (Cenozoic) are indicative of post-collision environments.

In each side of the Tabriz fault, the zoning is reversed relative to the other side and it could not be justified to be related to one subduction zone. It is more likely to belong to two separate subduction zones. It seems that in the Zagros region the subduction relates to southern Neo-Tethys, while in Lesser Caucasus-Arasbaran, this event relates to the northern Neo-Tethys, and these two events justify the magmatism and geotectonic evolution of the region. On the other hand, a northern ocean was existed between the north of Central Iran, south Armenia and Tauride blocks with Eurasian plate, and a southern ocean was between the above blocks with Arabian plate. The oceans were closed during convergence of the Arabian plate moving northward towards Eurasian plate. The dip direction of subduction in southern ocean is supposed to be northeast, while there are many controversies about the dip direction of the subducting slab of the northern ocean, although magmatic and petrogenetic characteristics of the area and the distribution of the magmatic rocks in time and space are indicative of probable southwest dip direction for subducting slab of the northern ocean.

Gold Mineralization in the SSZ were previously supposed to be related to orogenic events, while recently it is believed that they are intrusion related type mineralization that belongs to the Eocene post collision events. Recent investigations have also indicated the probable existence of porphyry type mineralization in the Hashtrud, a part of northwestern UDMB.

Carlin and epithermal type gold mineralizations are associated with Miocene adakitic magmatism, i.e. in Ghorveh and Takab region. Oligocene magmatism in the Arasbaran zone is associated with remarkable porphyry, skarn and vein type Cu-Mo-Au mineralization. Epithermal gold mineralization is accompanying both Oligocene and Miocene magmatisms.

In general, the northern belt including the AALCB is highly mineralized compare to the southern belt including the UDMB and SSZ that have relatively poor mineralization.

3. Acknowledgements

This study was supported by the Geological Survey of Iran (GSI). We acknowledge the faculty research grants from Miami University in support of fourth author work in eastern Turkey, Azerbaijan, and Northern Iran. Special thanks to Mr M.B. Dorri, N. Abedian, and B. Borna (GSI) for logistical support.

4. References

Aghanabati, A. (1993) Geological map of the Middle East, scale: 1:5,000,000. Geological Survey of Iran.

Aghazadeh, M., Castro, A., Rashidnejad Omran, N., Emami, M.H., Moinvaziri, H. and Badrzadeh, Z. (2010) The gabbro (shoshonitic)-monzonite-granodiorite association of Khankandi pluton, Alborz Mountains, NW Iran, Journal of Asian Earth Sciences, 38, 199-219.

Ahmadzadeh, Gh., Jahangiri, A., Lentz, D. and Mojtahedi, M. (2010) Petrogenesis of Plio-Quaternary post-collisional ultra-potassic volcanism in NW of Marand, NW Iran. Journal of Asian Earth Sciences, 39, 37–50.

Alavi, M., (2007) Structures of the Zagros fold-thrust belt in Iran. American Journal of Science, 307, 1064–1095.

Azizi, H. and Moinvaziri, H., (2009) Review of the tectonic setting of Cretaceous to Quaternary volcanism in northwestern Iran. Journal of Geodynamics, 47, 167–179.

Bazhenov, M., Burtman, V.S. and Levashova (1996) Lower and Middle Jurassic paleomagnetic results from the South Lesser Caucausus and the evolution of the Mesozoic Tethys Ocean. Earth and Planetary Science Letter, 141, 79-89.

Berberian M., Amidi, S.M. and Babakhani, A. (1981) Discovery of the Qaradagh ophiolite belt, the southern continuation of the Sevan-Akera (Caucasus) ophiolite belt in northern Iran (Ahar quadrangle), a preliminary field note, GSI.

Berberian, M. and King, G.C.P., (1981) towards a paleogeography and tectonic evolution of Iran. Canadian Journal of Earth Sciences, 18, 210–265.

Blourian, G.H. (1994) Petrology of the Tertiary volcanic rocks in the northern Tehran: [M.Sc. thesis], Tehran, Iran, University of Tarbiat Moallem, 145 pp.

Burnham, C.W. (1979) Magmas and hydrothermal fluids. In: Barnes, H.L. (Ed.), Geochemistry of Hydrothermal Ore Deposits. Wiley, New York, 71–136.

Dewey, J.F., Pitman, W., Ryan, W., and Bonin, J. (1973) Plate tectonics and the evolution of the Alpine system. Geological Society of America Bulletin, 84, 3137–3180.

Dilek, Y., Imamverdiyev, N. and Altunkaynak, S. (2010) Geochemistry and tectonics of Cenozoic volcanism in the Lesser Caucasus (Azerbaijan) and the peri-Arabian region: Collision-induced mantle dynamics and its magmatic fingerprint. International Geology Review, 52, 536-578.

Dilek, Y. and Furnes, H. (2009) Structure and geochemistry of Tethyan ophiolites and their petrogenesis in subduction rollback systems. Lithos: Doi:10.1016/j.Lithos.2009.04.022.

Galoyan, Gh., Rolland, Y., Sosson, M., Corcini, M., Billo, S., Verati, Ch. and Melkonyan, R. (2009) Geology, geochemistry and $^{40}Ar/^{39}Ar$ dating of Sevan ophiolites (Lesser Caucasus, Armenia): Evidence for Jurassic back-arc opening and hot spot event between the South Armenia block and Eurasia. Journal of Asian Earth Sciences, 34, 135-153.

Ghaderi, M., Ramazani, J. and Bowring, S. (2009) Ages of plutonic activity in the Sanandaj-Sirjan zone, Iran: Implication for plate convergence and Zagros collision. Geological Society of America Abstracts with Programs, 41, No. 7, p. 482.

Ghalamghash, J, Nedelec, A, Bellon, B, Vousoughi, Abedini, M, and Bouchez, J.L. (2009) The Urumieh plutonic complex (NW Iran): A record of the geodynamic evolution of the Sanandaj–Sirjan zone during Cretaceous times – Part I: Petrogenesis and K/Ar dating. Journal of Asian Earth Sciences, 35, 401–415

Harris, N.B.V., Pearce, J.A. and Tindle, A.G., (1986) Geochemical characteristics of collision zone magmatism. In: Coward M.P. and Ries, A.C. (Eds), Collision Tectonics. Geological Society of London, Special Publication. 19, 67-81.

Hessami, K., Koyi, H.A. and Talbot, C.J. (2001) The significance of strike slip faulting in the basement of the Zagros fold and thrust belt. Journal of Petroleum Geology, 24, 5–28.

Hezarkhani, A. (2006) Petrology of the intrusive rocks within the Sungun Porphyry Copper Deposit. Journal of Asian Earth Sciences, 26, 683–693.

Jahangiri A. (2007) Post-collisional Miocene adakitic volcanism in NW Iran: Geochemical and geodynamic implications, Journal of Asian Earth Science, 30, 433-447.

Jamali, H., Dilek, Y, Daliran, F., Yaghubpur, A. and Mehrabi, B. (2010) Metallogeny and tectonic evolution of the Cenozoic Ahar- Arasbaran volcanic belt, northern Iran. International Geology Review, 53, 608-630.

Karimzadeh, A., (2004) Marano volcanic rocks, East Azerbaijan province, Iran, and associated Fe mineralization. Journal of Asian Earth Sciences, 24, 11-23.

Keskin, M., (2003) Magma generation by slab steepening and break off beneath a subduction-accretion complex: An alternative model for collision-related volcanism in Eastern Anatolia. Geophysical Research Letters, 30, No. 24, 1-4.

Kheirkhah M., Allen M.B. and Emami M. (2009) Quaternary syn-collision magmatism from the Iran/Turkey borderlands, Journal of Volcanology and Geothermal Research, 182, 1-12.

Khezerlu, A.A., Amini, S. and Moayed, M. (2008) Petrology, geochemistry and mineralography of potassic to ultra-potassic (mainly shoshonitic) lavas of NW Marand (Azerbaijan). Journal of Science, Tarbiat Moallem University, 8, No. 3, 183-204.

Le Bas, M.J., Le Maitre, R.W., Streckeisen, A., and Zanettin, B. (1986) A chemical classification of volcanic rocks based on total alkali-silica content. Journal of Petrology, 27, 745-750.

Le Maitre R.W., Bateman P., Dudek A., Keller J., Lameyre Le Bas M.J., Sabine P.A., Schmid R., Sorensen H., Streckeisen A., Woolley A.R. and Zanettin B. (1989) A classification of igneous rocks and glossary of terms. Blackwell, Oxford.

Mazhari, S. A., Bea, F., Amini, S., Ghalamghash, J., Molina, J. F., Montero, P., Scarrow, J. H. and Williams, I. S. (2009) The Eocene bimodal Piranshahr massif of the Sanandaj–Sirjan Zone, NW Iran: a marker of the end of the collision in the Zagros orogen, Journal of the Geological Society, London, 166, 1–17. doi: 10.1144/0016-76492008-022.

Mehrabi, B., Mahmoudi, Sh., Masoudi, F. and Crfou, F. (2009) Mesozoic and Cenozoic U- Pb ages and magmatic history of granitoid bodies in the northern Sanandaj-Sirjan metamorphic zone, Iran. Geological Society of America, 41, No. 7, 481pp.

Mehrabi, B., Yardley, D. and Cann, R. (1999) Sediment-hosted disseminated gold mineralization at Zarshuran, NW Iran, Mineralium Deposita, 34, No: 7, 673-696.

Moayyed, M., (2001) Geochemistry and petrology of volcano-plutonic bodies inTarum area, Tabriz University, [Ph.D. thesis]: 256 pp. (in Persian).

Moinevaziri, H., Aziz, H., Mehrabi, B., Izadi, F. (2008) Oligocene magmatism in the Zagros thrust zone (Sahneh-Marivan axis): the Second period of the Neotethyan subduction in the Paleogene. Tehran Universit. J. Sci. (in Persian) 34 (1), 113-122.

Moritz, R., Ghazban, F., and Singer, B.S. (2006) Eocene gold ore formation at Muteh, Sanandaj–Sirjan tectonic zone, Western Iran: A result of late-stage extension and exhumation of metamorphic basement rocks within the Zagros Orogen. Economic Geology, 101, 1497–1524.

Omrani, J, Agard, Ph., Whitechurch, H, Benoit, M., Prouteau Gand Jolivet, L., (2008) Arc-magmatism and subduction history beneath the Zagros Mountains, Iran: A new report of adakites and geodynamic consequences. Lithos, 106, 380–398

Pearce, J.A., (1996) Sources and settings of granitic rocks. Episodes, 19, 120–125.

Pearce, J.A., Harris, N.B.W., and Tindel, A.J. (1984) Trace element discrimination diagram for the tectonic interpretation of granitic rocks. Journal of Petrology, 25, 956- 983.

Richards, J. P. (2009) Postsubdution porphyry Cu-Au and epithermal Au deposits: Products of remelting of subduction-modified lithosphere. Geological Society of America.

Richards, J. P., Wilkinson, D. and Ullrich, Th. (2006) Geology of the Sari Gunay epithermal gold deposit, Northwest Iran. Economic Geology, 101, No. 8, 1455- 1496.

Rickwood, P.C., (1989) Boundary lines within petrology diagrams which use oxide of major and minor elements. Lithos, 22, 247- 264.

Rolland, Y., Billo, S., Corsini, M., Sosson, M. and Galoyan, G., (2009) Blueschists of the Amassia-Stepanavan Suture Zone (Armenia): Linking Tethys subduction history from E-Turkey to W-Iran. International Journal of Earth Sciences, 98, 533–550.

Shengor, A.M.C. and Natal'in, B.A. (1996) Paleotectonics of Asia: Fragments of a synthesis, in Yin, A., and Harrison, T.M., eds., The Tectonic Evolution of Asia. Cambridge: Cambridge University Press, 486–640

Shengor, A.M.C. (1990) Anew model for the Late Paleozoic Mesozoic tectonic evolution of Iran and implications for Oman, in Robertson, A.H.F., Searle, M.P. and Ries, R.C., editors, The Geology and Tectonics of the Oman Region :London. Geological Society, Special Publication, 49, 797–831.

Sosson, M., Rolland, Y., Corcini, M., Danelian, T., Stephan, J.F., Avagyan, A., Melkonian, R., Jrbashian, R., Melikian, L. and Galoian, G. (2005) Tectonic evolution of Lesser Caucasus (Armenia) revisited in the light of new structural and stratigraphic results. Geophysical Research Abstracts, 7, 06224.

Takin, M. (1972) Iranian Geology and continental drift in the Middle East. Nature, 235, 147–150.

Talebian, M. and Jackson, J., (2004) Areappraisal of earthquake focal mechanisms and active shortening in the Zagros mountains of Iran. Geophysical Journal International, 156, 506–526.

Thieblemont, D. and Tegyey, M., (1994) Unediscrimination ge´ochimique desrochesdiffe´rencie´este´moindeladiversite´d´origineetdelasituationtectoniquedes magmas: Comptes Rendusdel' Acade´ miedessciences, Paris, 319, No.2, 87–94.

Whitney, J.A. (1975) Vapor generation in quartz monzonite magma: a synthetic model with application to porphyry copper deposits. Economic Geology, 70, 346–358.

Whitney, J.A. and Stormer, J.C. (1985) Mineralogy, petrology and magmatic conditions from the Fish Canyon Tuff, central San Juan volcanic field, Colorado. Journal of Petrology, 26, 726–762.

Wilson M. (1989) Igneous petrogenesis, Unwin Hyman, London.

Winchester, J.A. and Floyd, P.A. (1977) Geochemical discrimination of different magma series and their differentiation products using immobile elements. Chemical Geology, 20, 325–343.

Zakariadze, G.S., Dilek, Y., Adamia, Sh.A., Oberha"nsli, R.E., Karpenko, S.F., Bazylev, B.A. and Solov'eva, N. (2007) Geochemistry and geochronology of the Neoproterozoic Pan-African basement of the Transcaucasian Massif (Republic of Georgia) and implications for island arc evolution in the Late Precambrian Arabian-Nubian Shield. Gondwana Research, 11, 92–108, Doi: 10.1016/j.gr.2006.05.012.

5

Allchar Deposit in Republic of Macedonia – Petrology and Age Determination

Blazo Boev[1] and Rade Jelenkovic[2]

[1]Faculty of Natural and Technical Science, "Goce Delcev"
University – Stip

[2]Faculty of Mining and Geology, Belgrade University

[1]Republic of Macedonia

[2]Republic of Serbia

1. Introduction

The Allchar Sb-As-Tl-Au volcanogenic hydrothermal deposit is situated at the northwestern margins of Kožuf Mts. (Republic of Macedonia), close to the border between Republic of Macedonia and Greece (Fig.1). From the geotectonic point of view, ore mineralization is related to a Pliocene volcano-intrusive complex located between the rigid Pellagonian block in the west, and the labile Vardar zone in the east. From the metallogenic point of view, the Allchar deposit belongs to the Kožuf ore district as part of the Serbo-Macedonian metallogenetic province.

2. Geology of the Kozuf district

Geologically viewed the Kozuf district is built of several geologic formations distributed in several stratigraphic complexes: complex of Precambrian metamorphic rocks; complex of Paleozoic metamorphic rocks; complex of Triassic-Jurassic sedimentary rocks; complex of Upper-Cretaceous sedimentary rocks; complex of Upper-Eocene rocks; complex of Pliocene sediments and pyroclasts and complex of Quaternary sediments.

The geologic structure also includes magmatic rocks represented by: complex of metamorphosed rhyolites and pyroclasts; complex of serpentinized ultramafic rocks; complex of basic igneous rocks, and complex of volcanic of calc-alkaline suites.

The complex of Precambrian metamorphic rocks is built of albitic gneisses and marbles situated in the vicinity of the Mala Rupa metamorphic block on the east. On the west (Mount Kozuf) the complex is built of gneisses and micaschists located in the Elen Supe tectonic block.

Gneisses and marbles have been found in the tectonically emersion block at Mala Rupa, west of the village of Konsko. Rakicevic et.al, (1970) determined a Precambrian age.

Besides the metamorphic rocks in the Mala Rupa block in the western flank of the Vardar zone and the Pelagon (Kozjak Mt), a block of metamorphic rocks-Elen Supe containing rocks of different composition, but similar degree of metamorphism, was also determined. The

Elen Supe block is built of gneisses and micaschists and its composition is similar to that determined for the lower parts of the Pelagonian metamorphic complex.

The complex of Paleozoic metamorphic rocks, unlike the Precambrian gneisses and marbles, is of lower metamorphic degree. It conformably overlies marbles of Precambrian age.

Paleozoic metamorfic rocks are most common in Adzibarica, between Keci Kaj and Gladnica, Jelovarnik and Porta as a phyllite horizon, a horizon of phyllitic schists and cipolines, quartzporphyry, phyllites, argilloschists and metasandstones with marble interlations and finally a horizon of quartzites, quartz schists and metadiabases.

The phyllite horizon also contains sericitic and epidotic schists, cipolines and marbles and metamorphosed quartzporphyry intruded by quartz veins (Adzibarica, between Keci Kaj and Gladnica). Graphitic schists have also been noticed in the series of sericitic schists. The horizon is approximately 750 m thick. It was also revealed in Adzibarica, Jelovarnik and Porta as well as in the vicinity of the Dusnica River source.

Fig. 1. Geographical position of the Alsar deposit

Cipolines and marbles alternate phyllitic schists in the horizon of cipolines and phyllitic schists west of Flora, Alcak, Ursa and Jelovranik along the River Dosnica course. They are overlain by schistose quartz porphyry distinguished as a separate horizon.

A horizon of quartzites and metasandstones was determined in Boulska Reka near Dina, Kalugjerica and Usevica. It also contains meta diabases with sporadic sulphide mineralization.

Rakicevic and Pendzerkovski (1970) determined these metamorphic rocks as early Paleozoic. Mersier (1973) determined the age of this series and that of Porta as Jurassic based on the degree of metamoprhism and because it concordantly overlies the Mala Rupa-Tsena series which he determined as Triassic.

Upper Cretaceous limestones overlie the horizon of phyllites, argilloschists and metasandstones with intercalations of marbles in the upper course of the River Dosnica.
The complex of Triassic-Jurassic sedimentary rocks is subdivided into two facies near the village of Uma (Rakicevic and Pendzerkovski, 1970):
-   A facies of poly coloured shales with intercalations of limestones
-   A facies of limestones and dolomite limestones of Triassic age.
The Jurassic is present as a facies of slab-like stratified limestones and a facies of limestones and clayey schists, quartzites and cherts in the Dve Usi, Flora and Jelovarnik localities. The rocks in the River Boula valley are covered by a thick series of pyroclastic rocks and tuffs.
The complex of Upper Cretaceous rocks is present as a series of limestones and conglomerates that corresponds to the Barremian and Albian, as well as a series of limestones of Turronian age.
The series of limestones is present in the Cardak, Dudica and Gladnica localities and in the Rzanovo and Studena Voda where these sediments comprise the top part of the nickeliferous-iron ores. The stratified limestones in the lower parts consist of marls with residues of *Nerinea olisoponensis cf. optuca, O. Turonica* fauna. Temkova (1962) considers them to be of Turronian age. According to Maksimovic (1981) the top stratigraphic border is outlined by the transgression of Alb-Cennomanian and Cennomanian when the weathering crust was mostly destroyed and re-deposited as oolitic sedimentary iron and bauxite ores.
Large portion of the Kozuf district is occupied by massive limestones, particularly in the Cardak, Dudica and Gladnica areas where they transgressively overlie Paleozoic rocks. Limestones are rather broken and karstified and 400 to 600 m thick. This is the largest thickness determined for the Sennonian limestones found in the Republic of Macedonia. Poorly preserved rudists are discovered in them and based on that data the age of these limestones was determined as Sennonian. Because of the large fissure density and karsification they represent water collectors for the rich sources of the Rivers Stara and Zarnica. The limestones of the Dudica district are intensively hydrothermally altered and intruded by young subvolcanic rocks.
The complex of Upper Eocene sediments is present as basal conglomerates overlain by flysch sediments. The basal conglomerates near the villages of Kumanicevo, Dragozel, Gornikovo and Barovo are mainly built of marly and limestone pebbles. Gabbro pebbles, diabases and limestones predominate in the conglomerates between Krnjevo and Barovo. Conglomerates alternate limestones and marly limestones or marls. The Upper Eocene sediments near the Barovo and Krnjevo villages are 800 m thick.
The complex of Pliocene sediments and pyroclasts is widespread in the Kozuf district. Essentially lacustrine sediments are built of coarse-clastic sediments that overlie the basement of various geologic formations. They overlie Upper Eocenic sediments between Barovo and Krnjevo. They are present as large-grained conglomerates and clayey sandstone sediments (between the villages of Dolna Bosava and Krnjevo).
Gravel sediments have been determined in the basement of the tuffs near the village of Gorna Bosava in the valley of the River Nistaica above the village of Cemersko. There are sands and clayey sands with intercalations of sand-clays or clayey sediments over the series of conglomerates near Krnjevo which itself is overlain by clayey carbonate rocks.
Marls overlain by clayey sand and clayey carbonate sediments with large amounts of fossil residues, bones and fauna (of mammals) occur near the village of Barovo. The last skeletons of this fauna were found in the topmost level of these clastic lacustrine sediments in diatomaceous earth beneath volcanic sediments-tuffs near Stukovi Orai in the vicinity of the

village of Barovo (Garevski, 1960). The age of the sediments was determined as lower part of the Upper Pliocene based on the fauna (Izmailov, 1960). Radovanovic (1930) determined the age of these sediments as Pontian.

The Pliocene clastic sediments in the southern parts of the basin end with a travertine and lie immediately beneath the pyroclastic sediments (above the village of Boula).

Pyroclastic sedimentary rocks cover Pliocene lacustrine sediments in the south parts of the basin near Vitacevo and Gatenovo. In the southmost part they overlie the rocks of the northern slopes of Mount Kozuf and extend along the Macedonian-Greek border, south of the village of Mrezicko. In the north they extend close to the town of Kavardci and Dolni Disan (south of Negotino). The final tuffs and conglomerates can be seen in the vicinity of the village of Radnja. The volcanic sediments are from several meters up to several hundred meters thick.

A horizon of aglomerative tuffs overlain by a horizon of fine-grained volcanic ashes and glass occurs in close proximity to the Mokliski Monastery in the valley of the River Luda Mara over the clastic lacustrine sediments present as carbonate clayey material. The latest horizon of volcanic sediments consists of brecciated well banded volcanic tuffs-pyroclasts. The largest blocks of volcanic rocks were found in the north slopes of Mount Kozuf, beneath the volcanic craters and domes (above the village of Radnja and Bara, in the vicinity of Gladnica, Ametkova Glava and Konopiste).

The complex of Quaternary sediments: Large amounts of significant Quaternary terraces are found right of the River Konska. The layers are 20 to 30 m thick. They are large-grained sediments consisting of rounded fragments, built mainly of gneisses, marbles and quartz that formed from crystalline rocks from the vicinity of Mala Rupa and Keci Kaj. These terrace sediments were completely worked out - washed for gold during ancient period that is noticeable in the terrain.

Larger terrace sediments are not noticed along the valley of the River Dosnica because of the steep river sides. However, today's terrace layers can be found in the river bed. Traces of washed out river terraces can be seen in individual parts, particularly in the lower river course at the end of the cliff below the village of Dren.

Quaternary tuff sediments (20 meters thick) can be seen near the village of Sermenin in an area of approximately 200 to 300 meters in size.They are located in the mouth of the River Belica and the River Sermeninska. Similar tuffs can be seen in the River Boulska near Dina.

The River Bosava with its tributaries brings mainly volcanic material, because it passes through volcanic sedimentary series in its upper and middle courses.

Deluvial coarse-grained clastic sediments overlie the volcanic sediments near Vitacevo. Redeposited volcanic glass and ashes, known as pemza and pumice, as well as redeposited aglomerative tuffs can be noticed along the Mrezicko-Kavadarci road.

The complex of metamorphosed rhyolites and quartzporhyry is located in Paleozoic schists or Jurassic metasediments in Adzibarica and Gladnica (Mersier 1973). It is known that they are interstratified in phyllite horizon conditionally determined as Paleozoic without any stratigrafic data. Quartzporphyry near Bel Kamen and Dve Usi in the vicinity of the villages Konsko and Dudica have also been identified.

Quartzporphyries are greyish-white to green rocks and represent metamorphosed magmatic rocks, or rhyolites and pyroclasts developed, most probably, at the same time period as the sedimentary rocks that formed the phyllites. Such rocks have also been established in the terrains of neighbouring Greece in Kastaneri, south of the village of Uma.

Based on data reported by Mersier (1973) they are of Upper Jurassic age. Rakicevic and Pendzerkovski (1970) determined these rocks as early Paleozoic.

The complex of serpentinized ultramafic rocks is situated in the Studena Voda-Rzanovo-Kumanicevo zone. It represents a tectonic structure on which serpentinites along with Jurassic and Upper Cretaceous metasediments cover Paleozoic and Triassic metamorphic rocks.

Lateritic deposits of nickeliferous iron ore developed over the Rzanovo-Studena Voda zone and along with sediments of the top parts were dynamometamorphosed in conditions of prehnite pumpelite up to greenschist facies (Boev, 1982).

A large mass of serpentinized ultramafic rocks is located in the River Mrezicka above the village of Mrezicko. Serpentinized masses are also found near Alsar. The serpentinites near Mrezicko and wider are highly tectonized grading, in some parts, into serpentine and talcshists. Smaal chromite pods are known in the ultramaphic rocks.

Large masses of serpentinized ultramafic rocks are also located along the Rzanovo-Studena Voda tectonic zone. Detailed petrologic investigations determined these rocks as dunites and harzburgites. They are almost completely altered to serpentinites and only in some places relicts of fresh ultramafic rocks can be seen. Gabbro pegmatites and rodingites have also been found in the zone.

The complex of mafic igneous rocks is present as gabbro diabase complex that occupies the eastern and north-eastern parts of Kozuf district. Gabbros, diabases and spilites predominate in the complex. Minor intrusions of leucocratic granitic rocks quartzmonzonitic in composition like those in Gornicet were determined in the bordering parts between intrusive mafic rocks-gabbros and effusive rocks-spilites. Granite-porphyry dikes were also determined in the mafic rocks near Smokvica, Davidovo and Dren.

Quartzdiorites and granodiorites were identified in the north-west part of the gabbro diabase complex, near the village of Boula and Radnja as well as near Milovan and along the valley of the River Dosnica. Smaller pegmatitic lodes intersecting quartzdiorites or granodiorites were also found near Radnja Tajder (1939) carried out detailed petrologic investigations of the gabbros of this complex and determined the following major types: wehrlite, troctolite, olivine gabbro, gabbro-eucrite, uralite-gabbro, diorite and quartzdiorite, basalts and diabases.

The easternmost parts of the gabbro diabase complex on both sides of the River Vardar (from Demir Kapija to Udovo) and even further to Gevgelija are represented by diabases, spilites and keratophyre. Spilites are the most abundant among them.

Karamata (1973) gave the basic genetic assumptions related to this gabbro diabase complex. He reports that Dren-Boula gabbro diabase complex is a product of multi stage extrusions of large amounts of basaltic magma forming, first, the diabase spilite parties. Later, the prompted new masses intruded beneath the diabase crust (rarely intersecting it) forming new diabase spilite extrusions. The intruded igneous masses partiallly differentiated, but the tectonic processes and magma pulasations precluded magma differentiation.

3. Structural and volcanic characteristics of the Kozuf district

The Kozuf district is a large volcanic complex situated in the south of the Republic of Macedonia. It spreads in the area of Mount Kozuf.

According to the regional geologic setting of the Balkans, it is part of the Vardar zone. In the east the Kozuf district is limited by a fault zone which is the west border of the Demir Kapija

- Gevgelija gabbro diabase ophiolite massif. In the west it is bordered by a fault zone that separates the Pelagonian massif and the Vardar zone.

The location of this volcanic complex in the Kozuf-Kilkis transverse zone (Arsovski, 1962) and the intersection with the Vardar zone indicates a central type volcanism, activated on the tectonic intersection formed by the reactivated regional fault structures of Vardar strike (NW - SE to N - S) and the Kozuf - Kilkis (E - W) fault structure formed during the neotectonic period. This type of volcanism is characterized by ring-radial structures.

A schematic morphostructural map of the Kozuf district, was made using the anlaysis of satellite scanograms, aerophotos and geologic data obtained by field investigations (Fig.2). The neotectonic fault structures grouped into three systems.

A fault system of Vardar strike are reactivated fault structures, the oldest being those of NW - SE and the youngest of N - S orientation. Products of both incipient and major phases of volcanic activity are located along these faults. Intensive hydrothermal activity (in the area of Dudica and Alsar) of N - S strike took place affecting the products of incipient volcanic activity.

A system of faults of NE - SW to E - W strike. This system is relatively younger than the Vardar system manifesting recent seismic activity. The intersection between this fault system and faults of Vardar strike points to the younger and final volcanic activity in the Kozuf district.

Ring structures are represented by several morphologically negative shapes (that can be seen in scanograms) and a positive structure in the area of Dudica. The area of the most striking negative ring structure (Vasov Grad-Mrezicko-Topli Dol-Rozden-Alsar) is built mainly of volcanic material but it also includes Pliocene as well as Triassic and Cretaceous sedimentary material. This composition, the concentric shape and radial pattern of internal rupture structures, the type of drainage indicate that this large ring structure is a collaps caldera.

The Dudica positive ring structure can not be seen in scanograms, but it can clearly be defined by field investigations and analyses such as drainage system. Most probably the volcanic activity started in the area of this positive structure (Stara Mircevica). The products of initial volcanic activity are hydrothermally altered and covered by the products of later and final volcanic activity. The volcanic activity in the Kozuf district started in the Miocene and the isotopic age of rocks was determined as 12.1 m.y. (Troesch and Frantz, 1992)

The volcanic characteristics of the Kozuf district were determined by field investigations and analyses of plane photographs. The volcanic activities produced volcanic necks, frozen supply channels, large quantities of pyroclastitic material. Lava flows and development of typical volcanic domes have not been identified. This results from the nature of the magmatic activity and the composition of magma that gave the material for the rocks during the final phases of differentiation.

The magmatic activity included intermediary, occasionaly acidic, magma which was immobile and fairly rich in volatile components. This led to a rapid closure of supply channels resulting in a large explosive phase during volcanic activities. This is proved by the large presence of pyroclastic and epiclastic material such as lacustrine tuffs, conglomerates, volcanic glass and ashes. The large amounts of borone and fluorine in the volcanic rocks from Kozuf points to the existence of a long duration of emanation phase in the evolution of this volcanism.

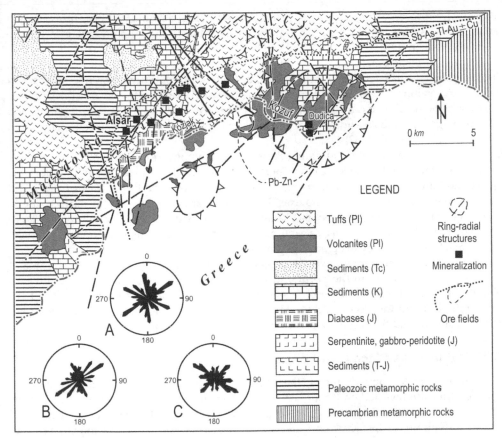

Fig. 2. Schematic morphostructural map of the Kozuf district (Boev, 1985)

4. Magmatism of the Kozuf district

The magmatic complex of the Kozuf district is a segment of the widespread magmatic activity in the petrographic province formed in the Vardar zone and the Serbo-Macedonian Mass.

General features of the petrographic province. - During the Tertiary, from the Eocene to the Pliocene, granodioritic magmas in this terrain intruded and extruded to the surface in individual tectonic zones. The evolution of this magmatism was first reported by Ilic (1962), and later shown in detail by Karamata (1962), and Karamata and Djordjevic (1980). The principal geochemical features of this magmatism were given by Karamata (1984) and Antonovic and Filipovic (1987), and individual areas were investigated in detail by Boev (1988) and Serafimovski (1990).

The Tertiary magmatism in the the Vardar zone and the Serbo-Macedonian Mass took place after closure of the Mesozoic oceanic basin (Karamata, 1983). This closure is due to the approach of the Dinaride slab and the Carpatho-Balkan block to the Serbo-Macedonian mass and the successive collison of the continental segments (Dimitrievic, 1974, Karamata, 1975,

1981). The process related to subduction during Dogger and Upper Jurassic was accompanied by calc-alkaline magmatism during the Middle and Upper Creataceous in the Carpatho-Balkanides.

Following-up continental collison resulted in thickening of the continental crust and its intrusion into the upper envelop and isostatic upliftings. Discontinuous compression caused temporary melting of the basal parts of the continental crust and mixing with larger or smaller amounts of material from the envelop (Knezevic et al., 1989). These pulsations and tactonomagmatic activities took place in the Oligocene, Miocene and Pliocene for several times (Thompson et al., 1982).

Magmas were distributed in individual areas, most commonly, in the middle parts of arch-dome structures and formed volcano-plutonic belts. Rocks formed from granodiorite, quartzdiorite to quartzmonzonite magmas and built intrusive bodies of various sizes and very large and small volcanic complexes. Rocks occur at and / or near the surface due to upliftings of individual tectonic blocks and the intensity of erosion. However, these complexes can be interpreted as volcano-plutonic complexes in which deep intrusive parts are sometimes revealed by deep erosion processes.

Generally viewed, these rocks occur in two belts that join in their middle parts (the area of Kopaonik) and separate to the north and northwest and southeast and south- southeast. These belts are not connected to one geologic unit, and are located on both sides of the ophioloite belt and intersect the geotectonic units of the Balkan Peninsula - the Dinarides, the Vardar zone and the Serbo-Macedonian mass under a slight angle.

5. Volcanic rocks of the Kozuf district

The volcanic rocks formed during the Pliocene along transverse tectonic structures of Vardar strike are revealed on Kozuf and Kozjak Mts. in the southern marginal parts of the Tikves - Mariovo Tertiary basin. Volcanic activity is manifested by the occurrence of numerous volcanic heaps which basically represent frozen supply channel, and large masses of pyroclastiic materials.Generally, the volcanic domes are distributed in a zone of east-northeast extension, most commonly on tectonic structures, in the places where they intersect older structures of northwest orientation (the Vardar strike). The transverse tectonic structures are of neotectonic age, formed in the Pliocene and lie parallel to the north margin of the Aegean valley between Thessaloniki and Kavala.

Volcanic activity in Mts. Kozuf and Kozjak is represented as various types of volcanic rocks and volcanoclasts (volcanic breccias, conglomerates and tuffs). Volcanoclasts occur as sedimentary layers in the southern parts of the Tikves-Mariovo Tertiary basin where they comprise the topmost parts of the sediments. In some places the volcanoclasts are 200 to 300 meters thick.

Volcanic rocks are present as alkali basalt (small bodies), quartzlatites (delenites), andesite-latites (trachyandesites), transitional latite-quartzlatite and quartzlatite-latite (delenite-latite), as well as latite, trachyte, trachyrhyolites and rhyolites.

The volcanic rocks of Kozuf and Kozjak Mts display greatest similarity to the series of volcanic rocks of the Buchim-Borov Dol ore district, both in their mineralogy and chemical compositions the only difference being in the time period of their formation. Namely, the rocks of Kozuf and Kozjak Mts. formed in the Pliocene, whereas those from Buchim- Borov Dol formed in the Upper Oligocene. The former are extrusive (and explosive), the latter are subvolcanic and subvolcanic to hypoabyssal facies which means that their individual upper parts are eroded deeper.

Petrology. *Alkali basalts* (trachybasalts) are the least abundant rocks in Kozuf district. They are established in the Bara locality near the source of the River Nisava. Similar rocks are found in the wider Tikves basin such as the marginal parts of the valley near the village of Koresnica, near Demir Kapija, Karaudzule on the Negotino-Stip road, near the villages of Debriste, Mrzen, Oraovec and Gaber north of Bojanciste (Tajder, 1940, Boev, 1988, Jankovic et, al,1977).

The basalt of Bara is a dark to black rock of porphyrytic texture. It is composed of andesine (with 42% An), amphibole, biotite and augite as phenocrysts and cryptocrystalline groundmass. Chemical analyses (Table 1) show that it is a basic rocks that contains SiO_2 ranging from 50.12 up to 51.20%, containing fairly large amount of MgO - the largest magnesium content among the volcanic rocks of Kozuf. It also contains some alkalis which classifies it as alkali basalt.

	1	2	3
SiO_2	50.12	50.75	51.20
TiO_2	0.65	0.58	0.60
Al_2O_3	16.70	15.86	17.80
Fe_2O_3	1.66	1.58	2.01
FeO	2.39	2.12	2.42
MnO	0.07	0.07	0.06
MgO	10.80	10.50	11.20
CaO	4.42	4.70	4.60
Na_2O	3.05	3.12	3.25
K_2O	3.51	3.45	3.65
P_2O_5	0.33	0.25	0.45
H_2O	6.37	6.50	5.72

Table 1. Chemical composition of alkali basalts of the Kozuf area (%)

Andesite porphyry volcanic rocks are established near Studena Voda, Tresten Kamen and Sreden Rid (Boev, 1988). They have pronounced porphyritic texture in which phenocrysts are represented by plagioclase that is consistent with basic andesine to acid labrador (about 50% An), amphibole, biotite and augite. The groundmass of the rock is microcrystalline, with vitrophyre base.

Chemical composition (Table 2) shows that they are intermediary rocks with 59.20 to 59.94% SiO_2 and that they have fairly large amount of Na_2O relative to K_2O, whereas the Al_2O_3 content ranges from 16.25 to 16.80%.

Latites and andesite-latites of Kozuf and Kozjak Mts. are porphyry volcanic rocks (calc-alkaline) composed of idiomoprhous phenocrysts of andesine (40 - 47% An), sanidine, amphibole, biotite and pyroxene. The groundmass is microcrystalline composed of microliths and plagioclases, sandinie, biotite and pyroxene. Apatite, ilmenite, rutile, pyrite and magnetite occur as accessory minerals. Chemical and geochemical analyses show that latites are intermediary rocks in which the SiO_2 content ranges from 58.67 to 60.86%, and that of Al_2O_3 from 17.38 to 18.20%. It should be mentioned that they have relatively uniform amounts of major oxides such as CaO, Na_2O, and K_2O that classifies these rocks as monzonites. The MgO content ranges from 1.11 to 2.43% which is a characteristic of calc-alkaline rocks (Table 3).

	1	2	3
SiO_2	59.94	59.75	59.20
TiO_2	0.54	0.56	0.60
Al_2O_3	16.30	16.25	16.80
Fe_2O_3	3.97	3.88	3.71
FeO	1.52	1.48	1.50
MnO	0.05	0.06	0.06
MgO	2.00	1.95	2.12
CaO	7.33	5.52	5.60
Na_2O	2.11	2.70	3.10
K_2O	0.83	0.85	0.92
P_2O_5	0.45	0.46	0.45
H_2O	3.60	6.35	5.75

Table 2. Chemical composition of andesites of Kozuf (%)

The distribution of microelements and rare earth elements is given in Fig. 8. The diagrams and data about the content of microelements and rare earths (Table 3) display that latites possess increased concentrations of incompatible LIL elements such as Rb, Ba, and Sr. The digrams (Fig.3) also indicate a pronounced minimum of europium that gives information about fractionation processes of primary magmas or the character of partial meltings. It is obvious that the rocks are fairly rich in light rare earths with respect to heavy rare earths, with amount of rare earths of 280 ppm.

Quartzlatites (delenites) are transition varieties to latites. They are the most widespread volcanic rocks in Kozuf. They have been discovered in Blatec, Golubec, Miajlovo, in the vicinity of Dudica (Cardak, Sarena a.a.), Porta, Bela Voda, up to typical quartzlatites (delenites) near Momina Cuka. This group of volcanic rocks contains all transition varieties from latites to quartzlatites and has leucocratic nature. Quartzlatites are rocks with porphyritic structure composed mainly of andesine phenocrysts (38 to 45% An) and sanidine. They also contain low amounts of femic minerals such as amphibole, biotite and augite. Individual types of quartzlatites such as those at Bela Voda, Cardak, Golubec etc. contain large-grained idiomorphic amphibole as well as more glass in the groundmass that gives the rocks dark-grey to black colour. Quartzlatites contain higher silicium dioxide content, almost equal content of alkali oxides and lower potassium oxide content than that in latites which gives the volcanic rocks (quartzlatite of Momina Cuka) more acidic nature (Table 4).

	1	2	3	4	5	6
SiO_2	60.86	58.67	59.97	59.68	60.37	60.04
TiO_2	0.52	0.71	0.62	0.65	0.62	0.62
Al_2O_3	18.20	17.81	17.65	17.38	17.53	17.61
Fe_2O_3	4.64	5.51	4.87	4.97	4.88	4.24
MnO	0.11	0.11	0.09	0.12	0.10	0.07
MgO	1.11	1.50	1.25	2.07	1.18	2.43
CaO	4.10	5.48	4.45	4.58	4.71	5.32
Na_2O	4.35	4.05	4.44	4.35	3.83	3.87
K_2O	4.75	4.71	4.99	4.76	4.94	4.18
P_2O_5	0.56	0.68	0.73	0.73	0.56	0.16
H_2O	0.80	0.78	0.92	0.72	1.28	1.17
Zn	100	80	100	100	90	90
Mo	1	2	1	2	1	1
Ni	20	30	30	20	20	30
Co	20	20	20	20	20	20
Cd	1	1	1	1	1	1
As	13	12	11	10	10	11
Sb	0.9	0.8	0.8	0.9	1	0.9
Se	0.2	0.2	0.1	0.3	0.2	0.1
Sc	10	15	11	12	10	11
Hf	5	6	5	5	5	5
Ta	0.8	0.8	0.7	0.6	0.8	0.9
Th	31	28	29	30	31	31
U	9	8	7	8	9	9
Rb	180	174	154	181	180	174
Zr	210	200	210	210	190	200
Sr	1170	1100	1110	1050	1120	1100
Ba	1760	1800	1850	1750	1850	1800
Cr	25	26	25	26	26	25
W	4	3	4	4	4	3
Cs	41	42	41	42	42	41
La	85	85	95	78	80	81
Ce	157	145	200	210	170	175
Sm	9.1	8.13	11.2	11.1	14.1	13.2
Eu	1.9	2.0	2.1	2.3	2.5	1.9
Tb	0.78	0.75	0.74	0.68	1.11	1.10
Yb	1.85	2.01	2.20	2.50	2.70	2.82
Lu	0.28	0.30	0.31	0.32	0.30	0.29

Table 3. Chemical composition (in %) and microelements (in ppm) of the latite and andesite-latite

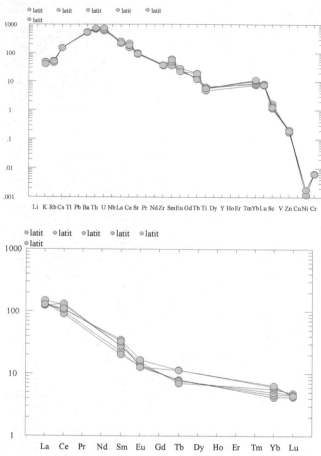

Fig. 3. Distributinot pattern of trace elements and rare earth elements in the latie and andesite-latiteof the Kozuf district (Jankovic et al, 1977)

The composition determined for quartzlatites classifies them in the alkali calcium group of rocks. Because of the large calcium and silica contents they are transitions between intermediary to acidic type of rocks. Their chemical composition is in agreement with their mineralogical composition since they are basically composed of plagioclases, potassium feldspars, amphibole and accessory minerals. It should be mentioned that taking in consideration the chemical composition of the rocks alone, would classify them as trachy-andesites or latites. However, from the aspect of their chemical composition, the presence of 14% of normative quartz in particular, the plagioclase and potassium feldspar ratio of 60:40, it is clear that they are quartzlatites.

Data related to the presence of microelements and REE indicates that the quartzlatites are enriched in LIL elements or the incompatible elements (Fig.4). They possess high contents of light elements, while total rare earths amount to 240 ppm. The rocks also contain fairly high arsenic and antimony amounts along with nickel and cobalt concentrations which is an indication of character of the deep fundament in the area.

	1	2	3	4	5	6
SiO_2	64.06	65.81	65.08	63.16	62.72	61.97
TiO_2	0.39	0.43	0.43	0.57	0.50	0.58
Al_2O_3	17.86	16.72	17.04	16.62	17.84	18.54
Fe_2O_3	3.02	2.90	3.39	4.44	4.12	3.82
MnO	0.03	0.05	0.08	0.09	0.08	0.07
MgO	1.44	0.61	0.47	1.32	0.79	0.52
CaO	3.69	3.12	5.04	4.20	3.64	2.40
Na_2O	4.21	4.56	4.34	3.92	4.09	4.74
K_2O	4.38	4.12	3.84	4.26	4.77	4.44
P_2O_5	0.19	0.39	0.54	0.50	0.54	0.19
H_2O	0.98	1.47	0.47	0.92	0.90	1.28
Zn	20	20	20	20	20	20
Mo	1	1	1	1	1	1
Ni	10	10	20	10	10	10
Co	10	10	10	10	10	10
Cd	1	1	1	1	1	1
As	10	10	10	10	10	10
Sb	0.8	0.7	0.8	0.7	0.8	0.8
Se	0.1	0.2	0.1	0.2	0.1	0.1
Sc	15	15	10	15	15	15
Hf	5	5	4	5	5	4
Ta	0.8	0.9	0.6	0.7	0.7	0.7
Th	27	28	28	29	28	27
U	7	8	8	7	6	7
Rb	190	210	200	180	190	210
Zr	220	210	220	220	210	220
Sr	1200	1250	1250	1200	1250	1250
Ba	1950	2000	2100	2100	1950	1900
Cr	20	20	20	20	20	20
W	3	4	4	3	4	5
Cs	40	41	39	39	40	40
La	62	65	66	63	63	67
Ce	140	138	115	120	125	125
Sm	7.3	7.4	6.8	7.1	7.2	7.2
Eu	1.52	1.50	1.38	1.47	1.42	1.54
Tb	0.7	0.7	0.7	0.7	0.7	0.7
Yb	2.0	1.6	1.7	1.8	1.8	1.8
Lu	0.30	0.39	0.38	0.34	0.34	0.35

Table 4. Chemical composition (in %) of quartzlatites and content of microelements (in ppm)

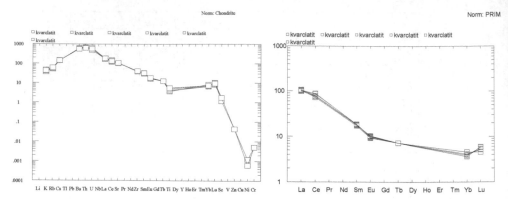

Fig. 4. Distributinot pattern of trace elements and rare earth elements in the qurtzlatites of the Kozuf district (Jankovic, et all, 1977)

Trachytes and trachyrhyolites are located in the westernmost parts of Kravica near the Sokol watch-house. The Kravica trachyte (an arsoite according to Tajder, 1940), occurs as a neck close to the Macedonian-Greek border in the territory of Greece. It is a well crystallized porphyry rock different in mineral composition from the rocks already described. It is composed of andesine, alkali feldspars such as sanidine and orthoclase and augite as femic mineral. The trachytes of the wider vicinity of Kravica are calc-alkaline in composition with large amounts of alkali oxides and higher potassium than sodium amounts that gives them pronounced potassic nature (Table 5). Chemical analyses indicate the presence of transition varieties called trachyrhyolites.

	1	2	3	4	5	6
SiO_2	55.82	55.81	55.52	58.39	56.16	60.12
TiO_2	0.95	0.86	0.92	0.93	0.93	0.55
Al_2O_3	18.41	18.06	18.88	19.17	17.76	17.84
Fe_2O_3	5.11	5.26	5.14	3.95	5.06	3.86
MnO	0.15	0.13	0.16	0.12	0.19	0.09
MgO	1.81	1.61	2.01	0.88	1.70	1.51
CaO	5.81	4.76	4.76	4.37	5.07	4.62
Na_2O	4.80	3.53	4.39	5.31	4.38	3.86
K_2O	5.74	6.50	6.37	6.10	6.26	5.05
P_2O_5	0.75	0.73	0.57	0.50	0.71	0.36
H_2O	1.09	2.26	2.22	1.15	1.38	1.27

Table 5. Chemical composition of trachytes (in %)

The Gradesnica Rhyolites are represented by lava extrusions of perlitic composition. Chemical analyses (Table 6) show that they are the most acid rocks occurring in the vicinity of Gradesnica west of Kozjak Mt. They are the last volcanic rocks formed in Kozuf and Kozjak Mts. They are of the Pleistocene (the Lower Quaternary) age and possess rhyolitic or vitrophyre composition. They are composed of glass with microliths of feldspars as small needles that have lava flow orientation. Large sanidine and plagioclase phenocrysts in their composition in some places make them typically porphyrytic. The rocks are fairly rich in silicium dioxide that gives them acidic nature. They are rich in alkalis, particularly potassium, but poor in calcium and magnesium oxides (Table 6).

	1	2	3	4	5	6
SiO_2	72.49	71.32	71.89	73.39	72.89	71.09
TiO_2	0.30	0.30	0.26	0.25	0.28	0.32
Al_2O_3	11.22	12.85	10.20	9.46	9.78	13.30
Fe_2O_3	6.19	4.95	6.61	8.04	8.04	4.13
MnO	0.12	0.12	0.12	0.11	0.15	0.26
MgO	0.14	0.22	0.93	0.37	0.25	0.18
CaO	0.78	0.75	0.55	0.40	0.60	0.71
Na_2O	2.87	3.21	2.15	2.32	2.46	3.24
K_2O	4.83	4.85	3.95	3.84	4.31	4.79
P_2O_5	0.06	0.60	0.08	0.03	0.07	0.03
H_2O	1.08	0.60	3.23	2.18	1.52	1.95

Table 6. Chemical composition of rhyolites of Kozuf (%)

6. Summary of the mineral composition of the volcanic rocks of Kozuf

The common feature of the volcanic rocks of Kozuf is the fairly high feldspar and almost equal amounts of plagioclases and potassium feldspar present as high potassium, calcium and sodium oxide in their mineral composition. Sporadically potassium content is higher than that of sodium oxides.

The disequilibrium in plagioclase and potassium contents results in the occurrence of transitional rocks types - from alkali to calc-alkaline series.

The silicium dioxide content results in the occurrence of transitional basic (from intermediary to acid) rocks that coincide with rhyolites. The SiO_2 content, including the basalt type of Bara (50.12%) ranges from 55.52% up to about 69%, only in exceptional cases to 73.39% SiO_2.

Locality	Rock type	SiO_2 content	Norm. quartz
Bara	basalt	50.12	-
Kravica	arsoit	56.12	-
Crna Tumba	trachyte	58.67	-
Dobro Pole	latite	60.86	5.73
Bela Voda	andesite-latite	60.04	8.50
Blatec	latite-Q latite	61.77	11.50
Dudica	quartz-latite	61.97	11.90
Momina Cuka	quartz-latite	63.68	12.00
Momina Cuka	quartz-latite	64.06	13.05
Mavra Petka	trachyte-riolite	66.44	16.74
Gradesnica	rhyolite	69.06	24.87
Gradesnica	rhiolite	73.39	41.26

Table 7. Correlation between SiO_2 content and normative quartz in the volcanic rocks of Kozuf

Quartz within phenocrysts can rarely be noticed in samples of volcanic rocks from Kozuf. It is mostly drawn into the groundmass and can not be seen under a microscope. Its content is shown as normative quartz calculated based on its chemical composition.

Another feature of the volcanic rocks from Kozuf is that they are characterized by the high feldspar abundances and the low amount of femic minerals such as pyroxene, biotite and amphibole in particular. It can be inferred that they are of salic nature because of 70% normative feldspars present in the rock. The total salic components amount to 95% (feldspars along with quartz). Table 7 gives the correlation between the SiO_2 and normative quartz in individual rock types from some localities.

Table 7 shows that the amount of normative quartz increases in rocks with higher SiO_2 content (the acidic igneous rocks).

An interesting analysis of the correlation between feldspars in individual rock types was also carried out (Table 8).

Locality	Rock	nor.fel.	nor.plag.	nor.K-fel	Pl / K f
Kravica	Arsoite	84.5	27.5	57	32 : 67
Dudica	Q latite	80.0	28	52	35 : 65
Tumba	Latite	81.0	43	38	53 : 47
M. Cuka	Q latite	80.5	46	34	57 : 43
B.Voda	and. Latite	75.5	44.5	31	59 : 41
Blatec	Latite	72.5	46	26.5	63 : 37
Bara	Basalt	67	49	18	73 : 27

Table 8. Correlations between normative feldspars in volcanic rocks of Kozuf

Data shown in Table 8 indicates that the amount of potassium feldspar increases with the increase of the amount of feldspar, but decreases with the decrease of the amount of feldspar. And vice versa, the amount of plagioclases decreases with the increase of the amount of feldspars, but increases with the decrease of the amount of feldspars. This behaviour is due either to the genesis of the rocks or the evolution of the primary isotopes.

7. Classification of the volcanic rocks of Kozuf

The diagram (Fig. 5) shows that only a small number of volcanic rocks analyzed belong to the field of andesites of subalkali nature. Most of the data related to the chemistry of the rocks plot in the field of latites and quartz latites (calc-alkaline rocks) with transition to trachytes (alkali rocks). The most acidic volcanic rocks plot in the field of rhyolites.

Fig. 6 shows classification of the volcanic rocks of Kozuf based on SiO_2 and K_2O contents. It shows that they contain high potassium contents. Only andesites plot in the field of rocks of low potassium.

The diagrams in Fig 7 and 8show that the volcanic rocks of Kozuf belong to the calc-alkaline series and that they are transitions between subalkali and alkali rocks.

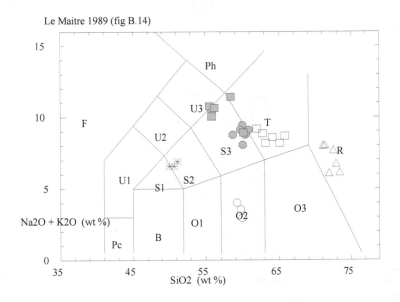

Fig. 5. Clasification of the Kozuf volcanic rocks based on the Le Maitre (1989) diagram (from Boev, 1988)

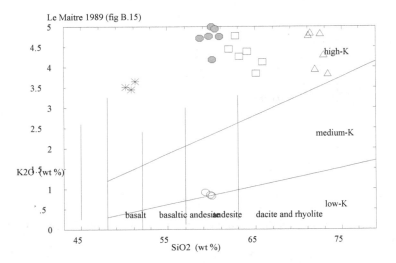

Fig. 6. Clasification of the Kozuf volcanic rocks based on SiO₂ / K₂O contents (Le Maitre, 1989) (from Boev, 1988)

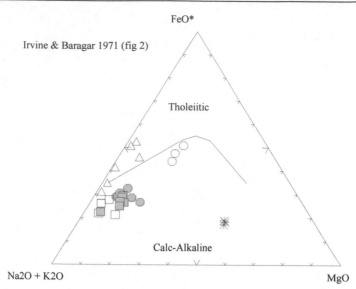

Fig. 7. Chemical composition of the Kozuf volcanic rocks (Irvine and Baragar, 1971) (from Boev, 1988)

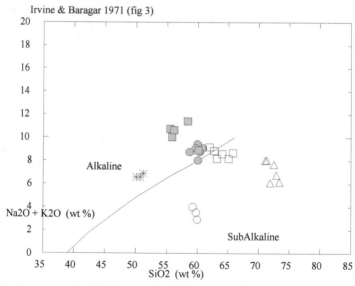

Fig. 8. Chemical composition of the Kozuf volcanic rocks (Irvine and Baragar, 1971) (from Boev, 1988)

8. Isotopic age of igneous rocks

Determination of isotopic age of Tertiary magmatism in the Republic of Macedonia was carried out by K/Ar method. Data obtained along with $^{87}Sr/^{86}Sr$ ratio are shown in Table 9.

Data indicate that the isotopic age in Tertiary volcanic and intrusive rocks ranges from 1.8 up to 29.0 m.y. depending on the locality. The youngest magmatic activity was determined for the Kozuf district (the south of Macedonia) where magmatic activity began in the period between the Miocene and Pliocene and terminated in the Quaternary. The highest isotopic value was determined for the rocks of Buchim-Borov Dol where the magmatic activity took place in the Oligocene.

The isotopic values for $^{87}Sr/^{86}Sr$ ratio indicate that they range from 0.706318 up to 0.710641.

Locality	Type of rock	Age in Ma	$^{87}Sr / ^{86}Sr$
Kozuf	Latite	1.8 ± 0.1	
Kozuf	Latite	5.0 ± 0.2	0.708546
Kozuf	Qurtzlatite	6.5 ± 0.2	0.709019
Kozuf	Andesite	4.8 ± 0.2	
Bucim	Latite	24.6 ± 2.0	0.706928
Borov Dol	Andesite	29.0 ± 3.0	0.706897
Damjan	Andesite	28.6 ± 0.6	0.706633
Zletovo	Quarzlatite	26.5 ± 2.0	0.706318
Zletovo	Latite	24.7 ± 0.4	
Zletovo	Monconite	21.9 ± 0.4	0.707770
Sasa	Andesite-latite	14.0 ± 3.0	0.710641
Sasa	Quartzlatite	24.0 ± 3.0	0.710244

Table 9. Isotopic age of Tertiary magmatic rock from Teritory of the Republic of Macedonia (Boev et al., 1991)

The first data related to the age of volcanic rocks from Kozuf were reported by Cvijic (1906). He discovered round pebbles of extrusive rocks in Neogene lacustrine sediments. He came to the conclusion that andesite in the area of Kozuf intruded Cretaceous limestones and that they are post Cretaceous in age, even older than the Tikves Neogene.

From investigations carried out on volcanic rocks of Kozuf, Radovanovic (1930) concluded that the products of the volcanic activity are lacustrine and coeval with the Neogene sediments.

Based on investigations carried out in the terrains of Greece, Kosmat (1924) reports of possible Pliocene age for the volcanic rocks of Kozuf.

Based on superposition relationships between the volcanic agglomerative tuffs and the Neogene lacustrine sediments in which pikermi fauna was discovered in the top parts Ivanov (1960) infers that, most probably, the rocks are of the Pontian age.

Based on investigations carried out on pollens Mersier and Sauvage (1965) infer that these rocks are of Pliocene age.

Measurements of the isotopic composition of the volcanic rocks on Voras Mt. (Kozuf) in Greece gave the following data (1.8 to 5.0 m.y)(Kolios et all, 1980).

Boev (1988) carried out measurements on the isotopic composition of volcanic rocks of Kozuf in the Republic of Macedonia. Results obtained are shown in Table 10.

Rock type and locality	K%	$K^{40}g/g$ x 10^{-6}	Ar %	$Ar^{40}cm^3$ x10^{-6}	$Ar^{40}g/g$ x10^{-9}	Ar^{40} /K^{40} g/g x10^{-3}	m.y.
Latite of Baltova Cuka	4.36	5.08	3.0 3.0	0.84 0.76	1.50 1.36	0.29 0.27	5.0 4.7
Latite-Quartzlatite of Vasov Grad	2.55	3.04	2.0 3.0	0.64 0.64	1.14 1.16	0.38 0.38	6.5 6.5

Table 10. Isotopic age of volcanic rocks of theKozuf district (Boev, 1988)

Lipolt and Fuhrman (1986) measured some volcanic materials as products of the hydrothermal zone of Alsar and obtained results as follow (Table 11):

Rock	mineral	K%	^{40}Ar (ccm/g) x 10^{-6}	^{40}Ar atm %	m.y.
Tuff	biotite	5.19	0.83	80.4	4.1±0.7
	feldspar	1.55	0.28	63.0	4.6±0.4
Tuff	biotite	7.04	1.21	52.2	4.4±0.4
	feldspar	5.90	1.01	51.7	4.4±0.5
	biotite	4.07	0.80	78.5	5.1±1.9
Andesite	feldspar	1.18	0.22	78.7	4.8±1.9
	ground mass	5.62	0.86	25.2	3.9±0.2

Table 11. Age of volcanics of the hydrothermal zone of Alshar (Lippolt and Fuhrman, 1986)

The following conclusions can be drawn based on data of isotopic investigations: the age of the rocks is in the span of 6.5 to 1.8 m.y. that determines a Pliocenic age. Individual differentiates are of Lower Pliocene age. Troesch, Frantz and Lepitkova (1995) report of data about the subvolcanic phase in Alsar that is in accord with the Miocene (12.1 m.y.).

9. Genesis of the volcanic rocks of the Kozuf district

Boev (1988) reports of some conclusions related to the origin of magmas that formed the volcanic rocks of the Kozuf district. He considered that magma sources were located in the marginal parts between the continental crust and the envelope. He gives data about isotopic $^{87}Sr/^{86}Sr$ ratio that supports this assumption.

Further investigations carried out by Lepitkova (1991) confirmed the earlier assumptions about origin of the magma that gave the material for the formation of the volcanic rocks. Namely, values determined for the isotopic $^{87}Sr/^{86}Sr$ ratio are within 0.708568 and are very close to those that Boev (1988) determined for the volcanic rocks of Kozuf.

Fig. 9.shows the relationship between the volcanic rocks of Kozuf and individual geotectonic areas in which magmatic processes took place. The diagram indicates that data on the volcanic rocks of Kozuf plot in the field of continental slab- like basalts or the so called within plate basalts.

This confirmation can be applied to explain the processes that contributed to the formation of the volcanic rocks in Kozuf.

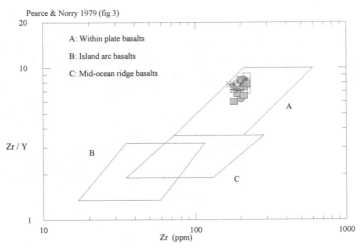

Fig. 9. Relations between the volcanic rocks of Kozuf and individual tectonic zones (based on Pearce and Norry, 1979) (From Boev, 1988)

It should be mentioned that CFB is related to the evolution of continental rift areas. The explanation about the genesis of magmas that gave the material for the formation of the volcanic rocks during the consolidation processes should be searched in the development of these structures present in the continental areas.

Chemistry of magmas related to continental rift zones is conditioned by heterogeneity of mineral and chemical compositions of the source in the envelop, the degree of partial melting and depth of its occurrence, the amount of magma that comes out on the surface etc. There is poor geophysical data related to the presence of magmatic sources in the upper parts of the petrographic provinces. This is important for the fractionation crystallization along with the evolution of the chemistry in magmas.

The genesis of the volcanic rocks of Kozuf can be explained best within the evolution of the Vardar zone as a rift zone, recurred several times during its evolution.

Based on available data related to the magmatic activity that took place in the Vardar zone from Oligocene to the Pleistocene it can be assumed that the processes of this geotectonic unit can be related to processes that took place in continental rift zones.

Favourable conditions for the formation of magmatic sources in areas of increased thermal activity is due to partial meltings that took place in the upper parts of the envelop which supplied more materials from the lower parts of the crust.

Primary magmas formed in this mode penetrated the surface along individual structures that formed as a result of the evolution of the area changing their composition by contamination and assimilation processes.

In addition, normalized values of distribution of rare earths (Fig. 10) are applied to explain the genesis of the volcanic rocks.

Fig. 10.displays that there is no pronounced anomaly of Eu or predominance in fractionation processes of primary magma materials. The content of rare earth elements, enrichment in light rare earths as well as the high content of LIL elements indicate that primary melt consisted of crystallized garnet that was conformable with residual plagioclase melt. Pressure in such systems amounts to some 15 Kb or 45 to 50 km depth.

From geophysical data of continental crust beneath Kozuf it can be inferred that the crust is about 40 km thick.

Based on the aforementioned data a conclusion can be drawn that magma sources were located in the marginal parts of the upper envelope and the lower crust taking in consideration, of course, erosional processes of several million years.

Norm: PRIM

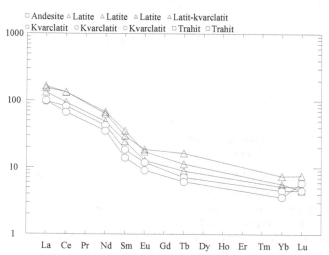

Fig. 10. The REE contents of the Kozuf volcanic rocks (From Boev, 1988)

The change in chemical composition of the extrusive magma yielding the series of volcanic rocks can be explained by assimilation processes that magma performed passing through different lithologic environments assimilating them but altering its composition as well. According to their mode of occurrence and spatial distribution the volcanic rocks of Kozuf are riftogenic, formed by magmatic activation in the marginal parts of the earth's crust and the upper envelop. This can be inferred from isotopic $^{87}Sr/^{86}Sr$ ratio that amounts from 0.7088 to 0.7090 (Boev, 1988).

Consolidation time, and the occurrence of volcanic rocks in the area were determined as 6 to 1.8 m.y. by K/Ar method (Boev, 1988) which is consistent with their stratigraphic age.

10. Principal ore seposits of the Kozuf distrct

The Kozuf metallogenic district is defined by its ore mineralization and controlling factors of its spatial distribution.

Based on available knowledge and the degree of investigations carried out several mineralization styles and metal associations such as copper, antimony, arsenic, thallium and gold were identified in the domain of the district. The Sb-As-Tl-Au is the predominating mineralization.

The following morphogenetic types are distinguished:

- Volcanogenic epithermal replacement mineralization of Sb-As-Tl association,
- Carlin-type gold mineralization,
- High sulphidation enargite vein-type mineralization accompanied, most probably, by gold mineralization as well
- Porphyry copper mineralization
- Epithermal Sb-As-Fe mineralization of vein-lense type related with fractures in the crystalline schists,
- Solfataric products are represented by native sulphur and marcasite, sporadically associated with galena.

The mineralization is of Pliocene age.

Magmatism of calc-alkaline suites and structures, both volcanic ring-radial and regional fractures are the principal factors controlling distribution of mineralization in time and space. Magmatic complexes are the principle source of ore metals and/or the source of heat energy for the formation of hydrothermal ore-bearing solutions that mobilized ore minerals from ultimate sources and transported them to the site of precipitation of mineral parageneses.

Fig. 2. shows the sites of mineralization of the Kozuf metallogenic district. The distribution pattern of mineralization is characterized by both lateral and vertical zoning. The central part of the district contains copper mineralization, surrounded by Sb-As ± Tl±Au deposit and occurrences (e.g. Alsar and Smrdliva Voda).

The Kozuf district is poorly explored, except of the Alsar deposit, and to some extent, the Dudica mineralization.

11. Alsar Sb-As-Tl-Au deposit

The Alsar complex Sb-As-Tl-Au deposit is one of the unique deposits in the world not because of its size, but mineral composition. It contains significant thallium concentrations that classify it as a unique deposit containing that metal. Besides economically significant antimony and arsenic concentrations, the Alsar deposit is the first Carlin- type gold deposit found in the Balkan Peninsula during the mid 1980s.

The latest mining activities started in 1881, and with some interruptions, lasted till 1913. During that period mainly arsenic ore was excavated and exported to Thessaloniki, Greece and Germany. Small amounts were mined out in the outcrops of the deposit. There is no data about the amount of arsenic ore mined out at that time.

The mineral potential of arsenic in the deposit is estimated at some 15.000 tons. According to today's criteria arsenic is a harmful component that results from antimony processing.

During the final years of the last century the first thallium minerals were discovered (lorandite, vrbaite a.a.) as constituents of arsenic-antimony ore.

Exploration for antimony carried out from 1953 to 1957 and from 1962 to 1965 resulted in the discovery of significant reserves of low grade ore. However, high arsenic contents in Sb-concentrations has precluded economic exploitation. The latest exploration for antimony was carried out in 1970-1973.

Mineral potential of the Alsar deposit, both mined out and available ore, exceeds 20.000 tons of antimony with 0.5% Sb as cut-off grade.

The name of the deposit pronounces as Alsar (Alshar), deriving the name of former Allchar mine (abbreviation of Allatini - a bank institution, owner of the concession, and Charteau - a mining engineer who worked in the mine).

Special interest for thallium as possible solar neutrino detector gave a new impulse for systematical investigations of thallium mineralization in the north part of the Alsar deposit (i.e. the Crven Dol ore body). This was an international LOREX Project aiming to establish reliability of the mineral lorandite from this deposit as thallium solar neutrino detector (Pavicevic, 1986, 1994). Some adits as no. 21 have been re- opened to enable taking the samples. This activity lasted from 1987 through 1993. Later it was restricted to laboratory investigations.

The mineral potential of thallium in the Alshar deposit has been estimated at 500 tons (order of magnitude).

The possible presence of gold in the Sb-As-Tl association at Alsar was initially suggested by Radtke and Dickinson (1984). During the 1986-1989 period gold mineralization was systematically explored. The results of both field and laboratory studies showed that the geological, geochemical, mineralogical and hydrothermal alteration features are strikingly similar to those which characterize Carlin-type mineralization of the Western United States (Percival and Radtke, 1990; Percival et al.,1992).

Unlike the Carlin-type gold deposits in the Western USA, the Alsar mineralization is hosted not only by sediments, but volcanics as well.

For the results of previous studies of the Alsar deposit, the reader is referred to Ivanov (1965, 1968), Jankovic (1960, 1979, 1982, 1988, 1993), Percival and Boev (1990), Percival et al. (1992), Percival and Radtke (1994), Boev and Serafimovski (1996) and for investigation of minerals to Krenner (1894), Locka (1904), Jezek (1912), Caye et al. (1967), Laurent et al. (1969), Johan et al. (1970, 1975), Terzic (1982), Balic-Zunic et al. (1986a,b, 1993), El Goresy and Pavicevic (1988), Palme et al. (1988), Jankovic and Jelenkovic (1994), Pasava et al. (1989, Frantz (1994), Frantz et al. (1994), Cvetkovic et al.(1994), Litbowitzky et al. 1995).

12. Main geological characteristics of the Allchar deposit

The Allchar deposit is polychronous and polygenetic. It has formed as a result of complex physico-chemical processes occurring in a heterogeneous geological environment, in the interaction of multi-stage hydrothermal fluids with the products of polyphase magmatic activity and surrounding sedimentary and metamorphic rocks. Mobilization, transportation and depositing of ore mineralization, as well as the supergene transformations of primary ore minerals, were determined by, and partly accompanied with, intensive pre-, sin-, and post-ore structural-tectonic terrain shaping. In those processes, there formed in the deposit several orebodies of varying ore shapes, textural-structural varieties and associations of minerals and elements, localized in various, tectonically predisposed geological environments.

The Allchar deposit is a NNW-SSE oriented antiform. It comprises several orebodies within a zone 2 km long and around 300-500 m wide. The localization of mineralization is spatially associated with an environments characterized by increased porosity and permeability, typically related to fractures and fractured zones in the vicinity of subvolcanic intrusive bodies. Such steeply dipping ore-bearing structures resulted from sliptype shearing movements represented by brecciated rocks often in a fine-grained gougy matrix. The increased porosity and composition of the tuffs are favorable environment for hydrothermal fluid migration and introduction of ore elements. Another favorable environment is a porous and permeable basal zone developed as a strata-bound body along the Triassic erosion surface (Percival, 1990; Percival and Radtke, 1994).

Mineralization is associated with hydrothermally altered wallrocks including the Triassic carbonates (dolomites and marbles), the Tertiary magmatic rocks and volcano-sedimentary sequence (tuffaceous dolomite). Silicification and argillitization are the most predominant alteration products, and quartz is very abundant in hydrothermally altered volcaniclastites (Percival and Radtke 1994; Pavićević et al. 2006). The alteration is generally believed to be associated with Plio-/Pleistocene andesite volcanism and latite intrusion, which extends from Kožuf Mts. in FYR Macedonia to Voras Mts. in Greece (Janković, 1993; Pe-Piper and Piper 2002; Yanev et al. 2008).

The major elemental components of the Allchar deposit are Sb, As, Tl, Fe and Au, accompanied by minor Hg and Ba, and traces of Pb, Zn, Cu. Enrichment of Tl in the Allchar deposit is closely associated with increased concentrations of volatiles, such as As, Sb, Hg.

The distribution of ore metals and their concentration rates display a lateral zoning. These zones are not sharply defined and typically a gradual transition exists between zones.

i.   In the northern part of deposit As and Tl prevail, accompanied by minor Sb, locally traces of Hg and Au.

ii.   The central part of deposit is dominated by Sb and Au, but also contains significant amounts of As, Tl, minor Ba, Hg and traces of Pb.

iii. The southern part of the deposit is characterized by dominance of gold mineralization accompanied by variable amounts of antimony.

The most important ore minerals of the Allchar deposit are Fe-sulphides, As- and Tl-minerals, cinnabarite, and Pb- and Sb-sulphosalts, accompanied by native gold or sometimes sulphur (Janković, 1960, 1993; Ivanov 1965; 1986; Janković and Jelenković 1994; Percival and Radtke 1994).

13. Local geologic setting

Deposition of sandstone and claystone, followed by bedded and massive carbonate rocks (limestone, dolomite, marble) took place in the Middle and Upper Triassic. These rocks are the basement of the Alsar deposit (Fig. 11).

The quartz-sericite-feldspar schists are developed along the eastern flank of the deposit, while the central part is built of dolomite, marble, and sporadically limestone.

The dolomite series underlies marbel. Based on fission traces the age of dolomite was determined as 250 m.y. (Lepitkova, 1995).

The Mesozoic rocks are unconformably overlain by Pliocene cover and glacial till. The earliest Tertiary rocks are very likely tuffaceous dolomite. It unconformable overlies the Mesozoic basement rock, particularly in the central, northern and southwest parts of the deposit. This unit is of volcano- sedimentary provenance and commonly mineralized.

The massive tuffaceous dolomite contains sporadically intercalated sequences of fine-grained tuff, water lain ash or volcanic glass. This volcano-sedimentary unit is 100-130 m thick (Percival, 1990). The basal contact of the tuff and underlying Tertiary tuffaceous dolomite and pre-Tertiary rocks is often marked by an unconformity zone, 2 to 12 m thick. It consists of a mixture of unsorted and ungraded detrital material. This basal unconformity of the tuff unit indicates a discontinuity in the Tertiary stratigraphic section and the beginning of volcanic activity during which dolomite deposition took place (Percival, 1990). The basal contact zone is of particular interest as a preferred environment of hydrothermal alteration and mineralization, particularly in the central and southern parts of the deposit.

The unit of Pliocene felsic tuffs covers a large portion of the Alshar deposit. This volcanic sequence includes ash, crystal tuffs, tuff breccia and lacustrine tuffaceous sediments.

According to Percival (1990) the lowest level of felsic tuffs consists of soft and friable ash tuffs, grading upwards to a crystal lithic tuff and then into a crystal tuff. These tuffs contain sanidine, biotite and quartz phenocrysts in an aphanatic ground mass.

The composition of tuff braccia is similar to the crystal tuff. The tuffs deposited in the sublacustrine basins in the southern part of Alshar show bedding and contain tuffaceous sedimentary clay material (a volcano-sedimentary series).

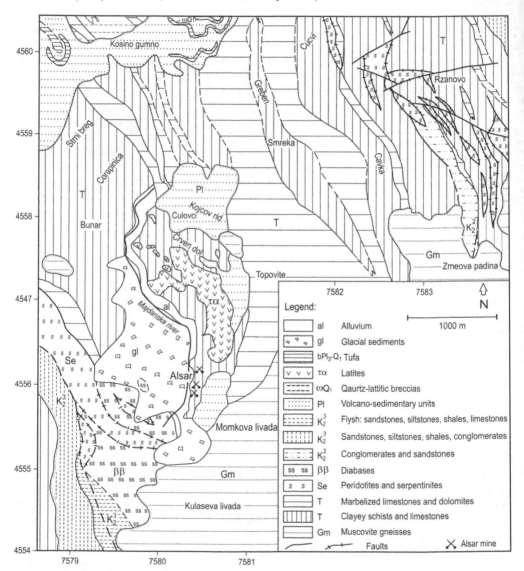

Fig. 11. Geological map of the Alsar area

The Alsar volcanic complex was investigated in detail by Frantz (1994), Frantz et al. (1994) and Lepitkova (1995).

Two principal volcano-intrusive phases have been identified in Alsar based on investigations carried out so far:

a.   a Miocene phase of calc-alkaline rocks ccurring as dikes. Troesch and Frantz (1994) have determined a Miocene age (14.3 - 8.2 m.y.) for the volcanic phase. The age was determined based on Ar/ Ar data obtained for plagioclase from Crven Dol (Table 12).

Minerals	Temperature (°C)	^{40}Ar / ^{39}Ar
Plagioklase	800	9.222 ±0.842
	1000	8.279 ±1.183
	1200	12.256±0.762
	1400	14.323±0.776

Table 12. Absolute age determination of volcanic rock from Alshar based on ^{40}Ar /^{39}Ar of plagioclase (Troesh and Frantz (1994)

The volcano-intrusive rocks of this volcanic phase were completely altered by hydrothermal solutes during the Pliocene.

b.   The most significant volcanic rocks in Alshar developed as part of the Kozuf volcano-intrusive activities. Subvolcanic hypoabyssal intrusions formed, based on data from K-Ar investigations, during the period from 4.5 to 5.0 m.y. (Lepitkova, 1995; Frantz et al. 1994).

Results obtained from determination of age by K/Ar method of andesines affected by hydrothermal processes indicate to absolute age of 3.9 to 5.1 m.y. (Lipolt and Fuhrman, 1986).

Determination of the age of volcanic rocks from Alshar was also carried out on sanidine and plagioclase (Table 13).

	Temperature (°C)	^{40}Ar / ^{39}Ar
Sanidinee	800	3.657 ± 0.137
	1000	3.334 ± 0.065
	1200	3.271 ± 0.063
	1400	3.261 ± 0.067
	1600	3.289 ± 0.170
Plagioklase	800	3.923 ± 0.319
	1000	3.328 ± 0.708
	1200	3.283 ± 0.757
	1400	5.027 ± 0.511
	1600	2.927 ± 0.058

Table 13. The age of volcanic rocks from Alshar determined by 40Ar / 39 Ar method (Troesh and Frantz, 1994)

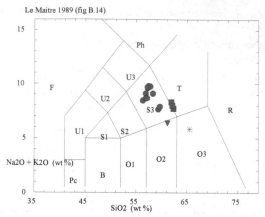

Fig. 12. Clasification of the Kozuf volcanic rocks based on the Le Maitre (1989)diagram (from Lepitkova, 1995)

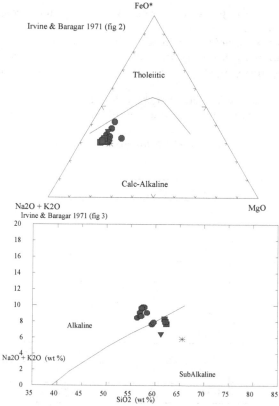

Fig. 13. Diagrams of (Na₂O+K₂O-FeO-MgO and Na₂O+K₂O/SiO₂) of volcanic rocks in the vicinity of Ashar (from Lepitkova, 1995)

It can be inferred that the volcanic activity in Alshar took place in the period between 3.9 to 5.1 m.y.. Based on Sr/ Sr ratio for latite (0.70856) it can be inferred that parent magma derived from lower continental crust/upper mantle domain (Boev, 1990/91).

Volcano-intrusive rocks of Alsar include latite, quartz-latite, trachyte, sporadically andesite and dacite (Fig. 12 and Fig.13).

The volcanics of Alshar contain variable amounts of trace elements and REE.Table 14 shows the values of trace elements and REE from three samples of trachytes (Frantz, 1994). Fig. 14

element	in ppm	error %	element	in ppm	error %
Li			Sn		
B			Sb	0.5270	10
C			Te		
F			J		
Na	167.0	3.0	Cs	1.330	6.0
P			Ba	34.80	13
S			La	6.340	3.0
Cl			Ce	7.300	5.0
K	41.00	15	Pr		
Sc	2.780	3.0	Nd	11.00	11
V			Sm	2.440	6.0
Cr	18.00	5.0	Eu	0.3400	6.0
Mn	38000	3.0	Gd	1.000	
Co	88.86	3.0	Tb	0.2400	13
Ni	< 20.00		Dy		
Cu	< 30.00		Ho	0.4700	20
Zn	593.0	4.0	Er		
Ga	< 0.6000		Tm	0.2700	20
Ge			Yb	1.200	6.0
As	637.8	3.0	Lu	0.1980	11
Se	< 0.40		Hf	0.2400	15
Br			Ta	0.03200	12
Rb	< 3.500		W	8.440	3
Sr	< 80.00		Re		
Y			Os		
Zr	10.0		Ir	< 0.00200	
Nb			Pt		
Mo	3.0	20	Au	0.494	3.0
Ru			Hg	0.7560	16
Rh			Tl		
Pd			Pb		
Ag	1.300	20	Bi		
Cd	12.00	20	Th	0.8000	14
In			U	8.520	5.0

Table 14. Trace elements of the trachytes from Alshar in ppm (Frantz, 1994)

shows the distribution of REE in the volcanic rocks of Alshar. It can be inferred that there is certain enrichment in light REE in regard to heavy elements. The relative enrichment in La is characteristic for the volcanic rocks of Alshar, whereas the Ce content (140-157 ppm), as well as Ce/Y (around 6) point to certain empoverishment in heavy elements. The Nd content is also high.

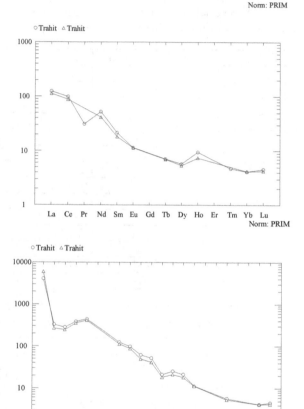

Fig. 14. Distribution pattern of REE (left) and trace elements of the trachytes from Alshar (right) (Frantz, 1994) (Lepitkova, 1995)

From the analysis it can be inferred that the enrichment in light REE elements indicates that magma originated from the continental crust and that it distinguishes it from toleitic basalts. The slightly pronounced minimum of Eu and the pronounced minimum of Dy indicate to the fractionation processes of primary magma and its contamination by rocks from the upper and lower crust (Lepitkova, 1995).

14. General characteristics of the volcanic rocks

Pliocene volcanic rocks occur either as subvolcanic intrusions in the shape of dikes and /or small intrusions, and extrusive volcanic material. Unlike Miocene volcanics, the younger rocks are relatively fresh and less affected by hydrothermal alteration processes (Lepitkova, 1995).

Latites.- Phenocrysts are clynopyroxene, andesine, sporadically biotite located in the ground mass of sanidine, andesine, Fe-Ti oxides and apatite.

The latites are the most common volcanic rock in the Alsar deposit. Their chemical compositions are displayed in Table 15.

	1	2	3	4	5	6	7	8	9	10
SiO_2	57.28	57.43	57.77	56.30	57.20	59.32	59.65	56.86	56.90	57.12
TiO_2	0.72	0.77	0.75	0.70	0.68	0.70	0.81	0.90	0.88	0.83
Al_2O_3	17.29	17.41	17.68	17.31	18.00	17.90	18.12	17.70	17.90	17.49
Fe_2O_3	5.60	5.84	5.73	5.20	5.62	5.25	5.38	5.30	5.32	5.48
MnO	0.06	0.08	0.06	0.06	0.06	0.10	0.10	0.10	0.10	0.09
MgO	1.89	1.60	2.00	3.21	1.80	1.66	1.78	1.88	1.98	2.01
CaO	4.42	4.23	5.35	5.18	4.68	5.07	4.87	5.12	4.95	4.88
Na_2O	4.01	4.10	4.31	3.65	4.15	3.65	3.79	4.01	3.95	3.88
K_2O	5.60	5.70	5.45	4.82	5.53	4.01	4.11	5.10	4.75	4.80
P_2O_5	0.57	0.51	0.50	0.44	0.51	0.52	0.50	0.48	0.44	0.47
H_2O	2.30	2.30	1.10	3.14	1.95	1.56	1.20	2.60	2.90	2.88
Total	99.74	99.97	100.7	100.01	100.18	99.74	100.31	100.05	100.01	99.93

Table 15. Chemical composition of the latites of Alsar (%)(from Lepitkova, 1995)

Trachyte is characterized by holocrystalline porphyry texture with phenocrysts represented by sanidine, plagioclase, amphibole and biotite. The groundmass consists of microliths of the same minerals.Chemical composition is presented on Table 16.

	1	2	3
SiO_2	61.90	62.30	62.08
TiO_2	0.70	0.65	0.72
Al_2O_3	17.80	18.01	17.23
Fe_2O_3	4.60	4.80	4.65
FeO			
MnO	0.10	0.09	0.09
MgO	1.30	1.39	1.35
CaO	4.72	4.30	4.18
Na_2O	4.01	3.61	3.85
K_2O	4.30	4.10	4.15
P_2O_5	0.50	0.50	0.50
H_2O	0.85	0.88	1.11
Total	100.78	100.63	100.63

Table 16. Chemical composition of the trachytes of Alsar (%) (Lepitkova, 1995)

Dacite and *andesite* occur sporadically. Their chemical composition is shown in Table 17.

	1 (andesite)	2 (dacite)
SiO_2	61.17	65.50
TiO_2	0.53	0.58
Al_2O_3	15.91	16.45
Fe_2O_3	4.68	3.43
FeO		
MnO	0.04	0.10
MgO	1.24	1.49
CaO	3.37	3.84
Na_2O	3.82	2.33
K_2O	0.24	3.50
P_2O_5		0.24
H_2O	4.77	6.79
Total	98.31	98.97

Table 17. Chemical composition of andesite and dacite of Alsar (%)(from Lepitkova, 1995)

Dacite consists of phenocrystas such as andesine, biotite, hornblende and minor quartz and sanidine, and of groundmass composed of microliths of andesine and sanidine as well as minor quartz, biotite, hornblende, apatite and Fe-Ti oxides.

Andesite is rare in the Alsar deposit. It is characterized by less alkalies than latite and trcahyte- andesite.

15. The age of mineralization and of volcanic rocks

No determination of age of ore minerals from the Allchar deposit using modern methods of laboratory testing has been ever conducted. Hence, the exact time of formation of orebodies from this deposit cannot be spoken of with any high degree of certainty. Based on the analysis of the geological setting of the Allchar deposit and its immediate surroundings, especially on the analysis of the control factors of the spatial position of the orebodies (magmatic, structural, and lithological control factors), as well as of other relevant geological parameters or ore mineralization (morphostructural and genetic types of orebodies, relation of orebodies to hydrothermal alterations, and so on), however, the formation of the deposit may be linked to the formation of the complex volcano-intrusive complex in the broader Allchar area.

The age of volcanism in the broader Allchar deposit area was determined on several occasions in the 1986-2009 period, using radiometric methods (K/Ar and Ar/Ar) on samples from various localities (Crven Dol, Kojčov Rid, Rudina, ADR and Vitačevo),

Determination of age using the K/Ar method was done on the following minerals, i.e., rocks from the Crven Dol locality: biotite and feldspars from tuffs (Lippolt and Furman, 1986); biotite, feldspars and pyroxenes from andesite (Lippolt and Furman, 1986), tuffs and volcanites (Boev, 1988), volcanites (Kolios et al., 1988), biotite from latite (Karamata et al., 1994).

Determination of age using the Ar/Ar method was done on minerals and rocks from the localities of Crven Dol, Kojčov Rid, ADR, Rudina and Vitačevo. The following were analyzed: sanidines from andesite from the Crven Dol locality (Troesh and Franc, 1992), biotites, feldspars and amphiboles from latite from the Kojčov Rid locality (Neubauer et al. 2009a), feldspars from the zones of intensively altered volcanites from the ADR and Rudina localities (Neubauer et al., 2009a), biotites from tuffs (Neubauer et al., 2009a)from the Vitačevo locality, and feldspars from tuffs from the Rudina locality (Neubauer et al., 2009a). The data on the petrologic and geochemical characteristics of the analyzed rocks may be found in the papers of Boev (1988), Yanev et al. (2008) and Neubauer et al. (2009a and 2009b), whereas the results of the conducted investigations are shown in Table 1.

Based on the above, it is possible to infer that two basic stages of volcanogenic-intrusive activity occurred in the broader Allchar deposit area: 1- the older, Miocene stage of volcanic activity, comprising the dikes of calc-alkali rocks fully hydrothermally altered during Pliocene and 2- a younger stage of volcanism of Pliocene age, which developed in the broader Allchar deposit area.

Judging by the results of investigations to date, the younger stage of volcanic activity occurred in several sub-stages in the period of ~6.5 do ~1.8 My (Boev, 1988). Andesites and tuffs from the Crven Dol area formed in the period of 6.5 to 3.9 My, tuffs from the Vitačevo and Rudina localities in the period of 5.1 ± 0.1 to 4.31 ± 0.2 My, according to Neubauer and coworker (Neubauer at al., 2009a), the biotite ages of 5.0 ± 0.1 and 5.1 ± 0.1 My from blocks of the Vitačevo tuff are geologically significant and interpreted to date the age of initial Pliocene volcanism in the Allchar region), whereas the latites from the Kojčev Rid locality formed in the period of 4.8 ± 0.2 to 3.3 My (Experiments with amphibole from a subvolcanic latite body result in disturbed $^{40}Ar/^{39}Ar$ release patterns and an age of 4.8 ± 0.2 My. Biotites yield slightly varying ages ranging between 4.6 ± 0.2 and 4.8 ± 0.2 My. K-feldspar disturbed, staircase patterns with ages increasing from 3.3 ± 0.2 to 4.0 My. The mineral ages of the subvolcanic latite body are interpreted, therefore, to monitor rapid cooling from ca. 550-500°C /amphibole/ through ca. 300°C /biotite/ to ca. 250 tu 160°C /K-feldspar/ between 4.8 ± 0.2 and 3.3 ± 0.2 My).

16. Conclusion

The geological setting of the Allchar deposit is dominated by the formations of the Middle and Upper Triassic age and by the calc-alkali volcanogenic-intrusive complex of the Neogene age. The prevalent lithological members in the central part of the deposit are carbonaceous rocks (limestones, dolomites and marbles), partly sandstones and claystones, whereas quartz-sericite-feldspar schists are prevalent in the eastern part of the deposit. Calc-alkali volcanism manifested in the occurrence of tuffs, lava flows, volcanogenic-sediment series and subvolcanic-hypabyssal intrusions with a genetic, and partly spatial relation to ore mineralization, is present in all parts of the deposit. The youngest rocks of Tertiary age are tuffaceous dolomites overlaying Mesozoic rocks in the central, northern and southwestern parts of the deposit. This unit is of volcanogenic-sedimentary origin and is frequently mineralized. In addition to those units, in the broader deposit area, there are also formations of Jurassic age (mafic-ultramafic complexes transformed into serpentinites), Neogene molasse sediments deposited in small lacustrine basins and Quaternary sediments. A characteristic feature of ore mineralization in the region of the Allchar deposit is the presence of zones in the spatial distribution of chief ore components (Sb, As, Tl, Au) and the accompanying associations of elements. In the northern part of the deposit (Crven Dol), there is a prevalence of As-Tl mineralization accompanied with Sb, locally and traces of Hg and Au. In the central part of the deposit, the basic ore components are Sb and Au, accompanied with As, Tl, minor Ba, Hg and traces of Pb (the central part); whereas the southern part of the deposit is characterized with mineralization of gold accompanied with varying concentrations of Sb and As.The age determinations of the volcanic rocks in the Alsar deposit indicate pliocene time.

17. References

Arsovski, M.,1962, Some Characteristics of the Tectonic Assembly of Central Part of Pelagonian Horst-antiklinorium and its Relations with Vardar Zone, Geological Survey of Macedonia, Book of Papers, 7, 37–63 (1962) (in Macedonian).

Balić-Žunić et al., 1986a: The crystal chemistry of thallium and its role in the mineralogy of Alšar. – In: The feasibility of solar neutrino detection with Pb205 by geochemical and accelerator mass spectroscopical measurements. GSI-86-9, Darmstadt, 39.

Balić-Žunić, T., Scavnicar, S., Engel, P., 1986b: The crystal structure of rebulite. Yt. f. Krist., 160, ½, 199

Balić-Žunić, T., Stafilov, T., Tibljas, D., 1993: Distribution of thallium and the genesis at Crven Dol locality in Alsar. Geologica Macedonica, 7, 1, 45-52, Stip.

Beran, A., M. Gotzinger, B. Rieck, 1990: Fluid Inclusion in Realgar from Allchar, in: Proceedings of the Symposium on Thallium Neutrino detection, Dubrovnik.

Boev, B., et al., 1982: Metamorphism of Ni-Fe ores from Rzanovo-Studena Voda and Zone Almopisa, Macedonica Geologica, No, 6, 24-31.

Boev, B., 1985: Petrological, geochemical and volcanic features of volcanic rocks of the Kozuf Mountain. PhD Thesis, Faculty of Mining and Geology, Stip, 195 pp (in Macedonian).

Boev, B., 1988: Petrological, geochemical and volcanic features of volcanic rocks of the Kozuf Mountain. PhD Thesis, Faculty of Mining and Geology, Stip, 195 pp (in Macedonian).

Boev, B., 1990/91: Petrological Features of the Volcanic Rocks from the Vicinity of Alshar. – Geologica Macedonica, 5, 1, Stip, 15-30.

Boev, B. et al., 1991: Oligocene-neogenic Magmatism in the Locality of the Bučim Block, Geologica Macedonica, 6, 23–32 (1991).

Boev, B. and Serafimovski, T., 1996: General Genetic Model of the Alshar Deposit, in: Proceeding of the Annual Meeting on the IGCP Project 356, Sofia, Vol. 1, 75-84

Caye, R., Picot, P., Pierrot, R., Permingeat, F., 1967: Nouveles donnees sur la vrbaite, sa teneur en mercure. – Bull. Soc. Franc. Min. Crist., 90: 185.

Cvetković, Lj.; Boronikhin, V. A.; Pavićević, M. K.; Krajnović, D.; Gržetić, I.; Libowitzky, E.; Giester, G.; Tillmanns, E. (1995). "Jankovićite, $Tl_5Sb_9(As, Sb)_4S_{22}$, a new TI-sulfosalt from Allchar, Macedonia". Mineralogy and Petrology 53: pp. 125-131.

Dimitrijević, M., 1974. The Serbo-Macedonian Massif. Tectonic of the Carpathian-Balkan Regions.

Geological Institute Dyoniz Stur, pp. 291–296

El Goresy A., Pavicevic, M.K., 1988: A new thallium mineral in the Alsar deposit in Yugoslavia. – Naturwiss. 75: 37-39, Springer-Verlag.

Frantz, E., 1994: Mineralogishe, geochemishe und isotopen-geochemiche Untersuchungen der As-Tl Sulfide in der Lagerstatte von Allchar. Doctoral Dissertation (Mainz, 1994).

Frantz, E., Palme, H., Todt, W., El Goresy, A., Pavićević, M.K., 1994: Geochemistry of Tl–As minerals and host rocks at Allchar (FYR Macedonia). Solar neutrino detection with Tl-205 the "LOREX" Project; geology, mineralogy and geochemistry of the Allchar deposit locality Crven Dol. N Jb Miner Abh 167: 359–399

Garevski, R., 1960, Neuer Fund von Mastodon in Diatomeen-schechten bei Barovo, Mazedonien: Skopje, Fragmenta Balcanica tome no. 16.

Irvine, T.N. and Baragar, W.R.A., 1971. A guide to the chemical classification of the common volcanic rocks. Canadian Journal of Earth Sciences, 8: 523-548.

Ivanov, T., 1960: Neue Angaben ubre den Vulkanismus im Kožuf Gebirge. – Bulletin scintifique, Tome, 5, No. 4.

Ivanov, T., 1965: Zonal distribution of elements and minerals in the deposit Alshar.- Symp. Problems of Postagmatic Ore Deposition, II, 186-191, Prague (in Russian).

Ivanov, T., 1986: Allchar the richest ore deposit of Tl in the world. GSI-report 86-9, Darmstadt, pp 6,

Izmailov N.A., 1960. Zhurn. fiz. khimii. 1960 T. 34. № 11. S. 2414

Janković, S., 1960: Allgemeine Charakteristika der Antimonit Erzlagerstätten Jugoslawiens. N. Jb. Mineral. Abh. 94: 506–538.

Jankovic, S., 1979: Antimony deposits of southeastern Europe. – Sesn. Zav. za geol. i geofiz. istraživanja, 37, 25-48, Belgrade.

Jankovic, S., 1982: Yugoslavia: in: Dunning et al., eds. Mineral Deposits of Europe, vol. 2. Southeast Europe, Mineral. Soc. and Inst. Min. Metal., London, 143-202.

Jankovic, S., 1988: The Alchar Tl-As-Sb deposit, Yugoslavia and its metallogenic features – in: Nuclear Instruments and Methods in Physic Research, A 271, 2, 286, North-Holland, Amsterdam.

Janković, S., 1993: Metallogenic features of the Alshar epithermal Sb-As-Tl deposit /The Serbo-Macedonian Metallogenic Province/. N. Jb. Miner. Abh., 166, 1, 25-41. Stuttgard.

Janković, S., Jelenković, R., 1994: Thallium mineralization in the Allchar Sb–As–Tl–Au deposit. N. Jb. Mineral. Abh., 167: 283–297

Janković, S., Boev, B., Serafimovski, T., 1997: Magmatism and tertiary mineralization of the Kozuf metallogenic district, the Republic of macedonia with particular reference to the Alsar deposit. Univ. "St. Kiril and Metodij" – Skopje, Faculty of Mining and Geology, Geological Department, Special Issue No. 5, 262.

Jezek, B., 1912: Vrbait, ein neues Thallium Mineral von Allchar in Macedonian. – Z. Krystallogr. 51: 3, 365-378.

Johan, Z., Pierrot, R., Schubel, H.J, Permigeat, F., 1970: La picotpaulit TlFe₂S₃, une nouvelle espece minerale. – Bull. Soc. fr. Miner. Cristallogr. 93: 544-549.

Johan, Z., Picot, P., Hak, J., Kvaček, M., 1975: La parapierrotite, un nouveau mineral thallifere d'Alchar (Yugoslavie). – Tschermarks Miner. Petrogr. Mitt. 22: 200-210.

Karamata, S., 1962: Tercijarni magmatizam Dinarida, njegove faze i njegove glavne petrohemijske karakteristike. (Der Tertiare Magmatismus in der Dinariden Jugoslawiens, seine Phasen und die wichtigsten petrochemischen Characteristiken) – V Savetovanje geologa Jugoslavije, Referati, II, 137-147 (in Serbian, Deutsch.), Beograd

Karamata, S., 1973: Les chromitites et leur relation genetique avec les roches ultramafiques de type alpin. – Colloque scientifique international E. Raguin: Les roches plutoniques dans leur rapport avec les gites mineraux, 397-398, Masson et Cie., Paris.

Karamata, S., 1975: Metallogenic provinces based on new results of geological researches. – IAGOD, Fourth symposium: Problems of ore deposition, Varna 1974, II, 486-492, Sofia.

Karamata and Djordjevic P., 1980: Origin of the Upper Cretaceous and Tertiary magmas in the Eastern Parts of Yugoslavia. – Bulletin LXXII de l'Acad. serbe des Sciences et des Arts, Classe des Sciences naturalles et mathematiques, 20, 99-108, Belgrade.

Karamata, S., 1981: Time and space in plate tectonic modelling of tectonic, magmatic and metamorhic processes in Tethys-type orogenic belts. – Bulletin LXXV de l'Acad. serbe des Sciences et des Arts, Classe des Sciences naturalles et mathematiques, 21, 27-46, Belgrade.

Karamata, S., 1983: Plate tectonic phenomena in the regions of the Tethys type. – Geotectonics, 5, 52-66, Moscow.

Karamata, S., 1984: Plate tectonic phenomena in the regions of the Tethys type. – Geotectonics, 17/5, Moscow.

Karamata S., Pavićević, M.K., Korikovskij, S.K., Boronihin, V.A., Amthauer, G., 1994: Petrology and mineralogy of Neogene volcanic rocks from the Allchar area, the FY Republic of Macedonia. – N. Jb. Minera. Abh., 167, 2/3, 317-328, Stuttgard.

Knežević, V., Steiger, R.H., Djordjević P., Karamata S., 1989: Precambrian contribution to Tertiary granitic melts at the southern margin of the Pannonian basin. – Symposium precambrian granitoids, Helsinki 1989, Abstracts, Geol. Survey of Finland, Spec. Papaer 8, 75, Eespoo.

Kolios, N., Innocenti, F., Maneti, P., Peccerillo, A., Guliano, O., 1980: The Pliocene volcanism of the Voras Mts. (Central Macedonian, Greece). – Bull. Volc. 43-3.

Kolios, N., Inocenti, F., Maneti, P., Pecerrillo, A., and Giuliani, O., 1980, The Pliocene volcanism of the Voras Mts.: Bulletin Volcanologique, v. 43, p. 553-568.

Kossmat, F., 1924: Geologie der centralen Balkanhalisen.

Krenner, J.A., 1894: Lorandite, ein neues Thalium Mineral von Allchar in Macedonian. – Math. es term. tud. Ertesio, 12: 473.

Laurent, Y., Picot, P., Pierrot, R., Permingeat, F., Ivanov, T., 1969: La raguinite TlFeS₂, une nouvelle espece minerale et le probleme de l'allcharite. – Bull. Soc. fr. Miner. Cristallogr. 92: 38-48.

Le Maitre, R.W., 1989. A Classification of Igneous Rocks and Glossary of Terms: Recommendations of the International Union of Geological Sciences Subcommission on the Systematics of igneous rocks. Blackwell, Oxford, 193 pp.

Lepitkova, S., 1991: Petrologic Features of the Volcanic Rock in the Vicinity of the Allchar Deposit with Particular Reference to Lead isotopes, M. Sc. Thesis, Faculty of Mining and Geology, Štip, University Ss. Cyril and Methodius, Skopje, 1995 (in Macedonian).

Lepitkova, S., 1995: Petrologic Features of the Volcanic Rocks in the Vicinity of the Alshar Deposit, with Particular Reference to Lead isotopes. – Master Degree Thesis, Faculty of Mining and Geology, Štip, 139 (in Macedonian).

Lippolt, H. J. and Fuhrmann, U., 1986. K-Ar age determination on volcanics of Alsar mine/Yugoslavia.-in: Proceed. Workshop on the feasibility of Solar Neutrino Detection with ²⁰⁶Pb by geochemical and mass spectroscopical measurements. – Nolte E., ed. Report GSI-86-9, Technische Univer. Munchen

Libowitzky, E., Giester, G. and Tillmans, E., 1995. The Crystal Structure of Jankovicite, $Tl_5Sb_9(As, Sb)_4S_{22}$.-Eur. J. Mineral., 7, 479-487.

Locka, J., 1904: Chemische Anayse des Lorandit von Allchar in Macedonien und des Clandetit von Szomolnok in Ungarn. – Z. Krist. Miner. 39: 520.

Maksimović, Z., Panto, GY., 1981: Nickel-bearing phlogopite fromthe nickel-iron deposit Studena Voda (Macedonia). Bulletin ofthe Serbian Academy of Sciences 80, 1–6.

Mercier, J. and Sauvage, J., 1965. Sur la geologie de la Mace'doine centrale: les tufs volcaniques et les formations volcano-detritiques plioènes á pollens et spores d'Almopias (Grèce). Ann. Geol. Pays Helléniques. 16, 188.

Mercier, J., 1973. Etude geologique des zones internes des Hellenides en Macedoine Centrale (Grèce): Ann. Géol. Pays Helléniques. 20, 1-798.

Neubauer, F., Pavićević, M.K., Genser, J., Jelenković, R., Boev, B., Amthauer, G., 2009: $^{40}Ar/^{39}Ar$ dating of geological events of the Allchar deposit and its host rock. FWF Report of the Project No 20594.

Peytcheva, I, von Quadt, A, Neubauer, F, Frank, M, Nedialkov, R, Heinrich, C and Strashimirov, St (2009). U-Pb dating, Hf-isotope characteristics and trace-REE-patterns of zircons from Medet porphyry copper deposit, Bulgaria: implications for timing, duration and sources of ore-bearing magmatism. Mineralogy and Petrology, 96, 19-41

Palme, H. et al., 1988: Major and trace elements in some minerals from Crven Dol, Allchar. Nucl. Instr. and Methods, 271 (2): 314-319.

Pašava, J., Petlik, F., Stumpfl, E.F., Zemman, J., 1989: Bertrandite, a new thallium arsenic from Allchar, Macedonia, with a determination of the crystal structure. – Miner. Mag. 53: 531-538.

Pavićević, M.K., 1986: Lorandite from Alshar as solar neutrino geochemical detector. – Proc. "The feasibility of the solar neutrino detection with Pb[205] by geochemical and accelerator mass spectroscopical measurements". – GSI-86-9, Darmstadt, 9.

Pavićević, M.K., El Goresy, A. 1988. Crven Dol Tl deposit in Allchar: Mineralogical investigation, chemical composition of Tl minerals and genetic implicatons. N.Jb.Miner.Abh., 167: 297–300.

Pavićević, M.K. & Amthauer, G. 1994. Solar neutrino detection with Tl-205 – The "LOREX" Project. I. Solar neutrino experiments, thallium neutrino detection and background reduction. II. Geology, mineralogy and geochemistry of the Allchar deposit locality Crven Dol. In: Pavićević MK, Amthauer G (eds) Proceedings of the International Symposium on Solar Neutrino Detection with Tl-205. October, 9–12, 1990, Dubrovnik, Yugoslavia. Beih N.Jb.Miner.Abh., 167 [Suppl]: 125–426

Pavićević, M.K, Cvetković, V., Amthauer, G., Bieniok, A., Boev, B., Brandstätter, F., Götzinger, M., Jelenković, R., Prelević, D., Prohaska, T., 2006: Quartz from Allchar as monitor for cosmogenic [26]Al: Geochemical and petrogenetic constraints. Mineral Petrol 88: 527-550

Pearce, J.A. and Norry, M.J., 1979. Petrogenetic Implications of Ti, Zr, Y, and Nb Variations in Volcanic Rocks. Contributions to Mineralogy and Petrology, 69: 33-47.

Pe-Piper G, Piper DJW (2002) The igneous rocks of Greece. The anatomy of an orogen. Gebrüder Borntraeger, Berlin – Stuttgart, 573 p

Percival, T.J., Boev, B., 1990: As-Tl-Sb-Hg-Au-Ba mineralization, Alšar district, Yugoslavia: A unique type of Yugoslavian ore deposit. – Int. Symp. on Solar Neutrino Detection with Tl[205]. Yug. Soc. Nucl. Elem. Part. Phys., Dubrovnik, 36-37 (abstract).

Percival, T.J., Radtke, A.S., 1990: Carlin Type Gold Mineralization in the Alsar District, Macedonia, Yugoslavia, in: Proceeding of the Eight Quadrennial IAGOD Symposium, Otawa, Canada, Program with Abstracts, P.A., 108.

Percival, T.J., Radtke, A.S., Jankovic, S., Dickinson, F., 1992: Gold Mineralization of the Carlin-type in the Alsar district, SR Macedonia, Yugoslavia. - Y.T. Maurice, ed. Proc. the IAGOD symp., Ottawa, Canada Proceed., E. Schweizerbaritsche Verlag, 637-646, Stuttgart.

Percival, J.C., Radtke, A.S., 1994: Sedimentary-rock-hosted disseminated gold mineralization in the Alsar District, Macedonia. Canad Mineralogist 32: 649-665

Radtke, A.S., Dickinson, P.W., 1984: Genesis and Vertical position of Fine-Grained Disseminated replacement Type Gold deposits in Nevada and Utah, USA. - In: Problems of Ore Depositio, Fourth IAGOD Symp. (Varna), 1, 71-78.

Rakičević, T., et all, 1970: Explanatory Note of the General Geological Ma   Yugoslavia, Sheet Kozuf, Federal Geological Survey of Yugoslavia, 1970 (in Macedonian).

Serafimovski, T., 1990: Isotopic Composition of the Sulphur in the Sulphides from Alshar. Geologica Macedonica, 5, 1, 165-172.

Tajder, M., 1939. Fiziografija, kemijski sastav i geneza gabroidskog masiva Dren-Boula u Južnoj Srbiji. Rad Jugoslavenske akademije znanosti i umjetnosti. Matematičko-prirodoslovni razred. Knj. 82

Tajder, M., 1940. Arsoit sa Kravičkoga Kamena i latit sa Tumbe. Rad Jugoslavenske akademije znanosti i umjetnosti. Matematičko-prirodoslovni razred. Knj. 83

Temkova V., 1962: Contribution à la connaisssance des sediments du Crétacé supérieur dans les environs du

village de Banjica (T. Veles).‡ Trud. geol. Zavoda Nar. Repub. Makedon., 9, 105‡119, Skopje. (In Macedonian, French summary).

Terzić, S., 1982: Thallium and mercury in As-Sb and Pb-Zn mineral paragenesis of Yugoslavia. - Gl. prir. muz., A 37: 51-115, Beograd (in Serbian with English summary).

Thompson, J,. 8., Jr., Laird, J. and Thompson, A, B. (1982) Reactions in amphibolite, greenschist and blueschist,Journal of Petrology, 23,1-27

Troesch, M., Frantz, E., 1992: ^{40}Ar/^{39}Ar Alter der Tl-As Mine von Crven Dol, Allchar (Macedonia). Eur J Mineral 4: 276.

40Ar/39Ar Alter der Tl-As Mine von Crven Dol, Allchar (Makedonien).-Beih. z. eur. J. Mineral., 4, No. 1, 276

Volkov, A.V., Serafimovski, T., Kochneva, N.T., Thomson, I.N., Tasev, G., 2006: The Alshar epithermal Au-As-Sb-Tl deposit, southern Macedonia. Geology of Ore Deposits 48:175-192

Yanev, Y., Boev, B., Doglioni C, Innocenti F, Manetti P, Pecskay Z, Tonarini S, D'Orazio M., 2008: Late Miocene to Pleistocene potassic volcanism in the Republic of Macedonia. Miner Petrol 94:4 5-60.

6

Late to Post-Orogenic Brasiliano-Pan-African Volcano-Sedimentary Basins in the Dom Feliciano Belt, Southernmost Brazil

Delia del Pilar Montecinos de Almeida[1],
Farid Chemale Jr.[2] and Adriane Machado[3]
[1]Universidade Federal do Pampa (UNIPAMPA)
[2]Universidade de Brasília (UnB)
[3]Centro de Geofísica da Universidade de Coimbra (CGUC)
[1,2]Brazil
[3]Portugal

1. Introduction

The Neoproterozoic/Early Paleozoic transition in Southern Brazil was marked by the last phases of the Brasiliano/Pan-African Cycle (900-540 Ma, Chemale Jr., 2000). These include the development of a trough series related to the different Brasiliano Orogenesis phases, which occupied one preferential subsiding locus called Camaquã Basin (Paim et al., 2000). This basin is located in the Sul-Riograndense Shield (Rio Grande do Sul State, Southernmost Brazil, Fig. 2). The Camaquã Basin is a northeast-southwest elongated basin positioned along the Dom Feliciano Belt limits (to the east) and the Rio de la Plata Craton (to the west). The basin was filled with a thick (ca. 11 km) sedimentary rocks package deposited in marine (at the base) and continental (at the top) environments, and a volcanic rocks series interbedded with the sediments. At least three volcanic events were recognized and related to the Hilário and Acampamento Velho Formations, and Rodeio Velho Member, from the base to the top (Fig. 1).

Paim et al. (2000) divided the Camaquã Basin into four sub-basins named as Taquarembó-Ramada, Santa Bárbara, Guaritas and Piquiri-Arroio Boici, from West to East (Fig.1). Each sub-basin showed a partial independent evolution, and consequently, the stratigraphic framework shows some differences in a regional scale. Santa Bárbara and Guaritas sub-basins show the most complete stratigraphic section. Paim et al. (2000) defined five major depositional cycles from tectonic origin separated by unconformities. Every cycle corresponds to a unit ranking as group or allogroup. From the base to the top, there are the following groups: Maricá, Bom Jardim, Cerro do Bugio, Santa Bárbara and Guaritas (Table 1).

This chapter aims to make the characterization of the two last volcanic events associated to the filling of Camaquã Basin, named as Acampamento Velho Formation and Rodeio Velho Member in Rio Grande do Sul, Brazil. The book chapter will be composed of an introduction. After the introduction, it will be presented the Camaquã Basin stratigraphy and evolution compilation, aiming to introduce the reader to the volcanic events characterization. We also present a compilation of geochronological data from different units that fill the Camaquã Basin as well as the U-Pb ages obtained by us for Rodeio Velho, Acampamento Velho, Hilário volcanic events and Maricá Formation (in sandstone). The mineralogy, petrography, geochemistry (major, trace and rare-earth elements) and isotope data will be presented to the Acampamento Velho Formation and Rodeio Velho Member.

2. Camaquã Basin stratigraphy and evolution

2.1 Camaquã Basin tectonic evolution

Camaquã Basin is related to a system of late to post tectonic basins associated with the Pan African Brasiliano Orogeny. In the late stages of the Brasiliano Orogenic phase (700-500 Ma); the depressions formed received sediments from the erosion of mountain areas still undergoing active uplifting (Almeida, 1969). Almeida (1969) and Almeida et al. (1976, 1981) subdivided the deposits of these basins in lower and upper molasse. The lower molasse sequence is folded and associated with volcanism, whereas the upper molasse is horizontal or tilted, associated with magmatism. Almeida (1969) attributed the configuration of these molassic basins to reactivation processes of the deep faults. Wernick et al. (1978) related their genesis to strike-slip movements. Other authors suggested that these basins were formed in depressions generated by synforms developed during the Brasiliano Orogenic event (Fragoso Cesar et al., 1982). According to Fragoso Cesar et al. (2000), Camaquã Basin is a Neoproterozoic – Early Cambrian extensional, post orogenic and pre-cratonic rift system. Paim et al. (2000) interpreted this basin as a depositional locus in which different basins succeeded one another, each of them with their own lithological record and distinctive subsidence mechanisms. These authors admitted a connection between the evolution of the Camaquã Basin and the final tectonic stages of the Pan-African Brasiliano Orogeny in Southern Brazil. They also mentioned the continentalization that took place in later stages, even as the clear tectonic control in the stratigraphic arrangement of the allogroups, thus suggesting, a residual tendency of crustal uplifting during the final stages of the Pan-African-Brasiliano Orogeny.

According to Chemale Jr. (2000), the subsidence of the Neoproterozoic Rio de la Plata Plate beneath the Kalahari Plate led to the formation of oceanic crust around 700 Ma. Rifting of the Kalahari Plate resulted in the formation of the Encantadas Microcontinent (or microplate), with a converging margin to the west and a diverging one to the east. The proto-ocean Adamastor was formed between 700 and 650 Ma. Between 650 and 540 Ma, it started to subside to the west, beneath the Encantadas Microplate. The calc-alkaline orogenic magmatism occurred between 630-610 Ma, and the later to post orogenic around 600-540 Ma (Phase III). The closing of this proto-ocean was achieved initially through transgression, changing to transcurrence in the final stages. The last deformational pulses related to the Brasiliano Event (Dom Feliciano) took place around 540 Ma. The deposition in the Camaquã Basin formed in the retroarc region, initially preceded through the accumulation of sedimentary sequences and associated to the magmatism at 620 Ma. There is evidence for progressive continentalization and alkaline signatures toward the top of the rock record about 500 Ma, when the final accretion of west Gondwana took place. Thus, these events marked the beginning of the stabilization of Gondwana Supercontinent and the initiation of erosional processes in the Brasiliano mountain ranges. Subsidence of the Paraná Basin occurred from 500 Ma onwards, with the accumulation of basal sediments in a rift environment. In the Sul-Riograndense Shield, this is probably represented by the Guaritas Group (Fig. 2).

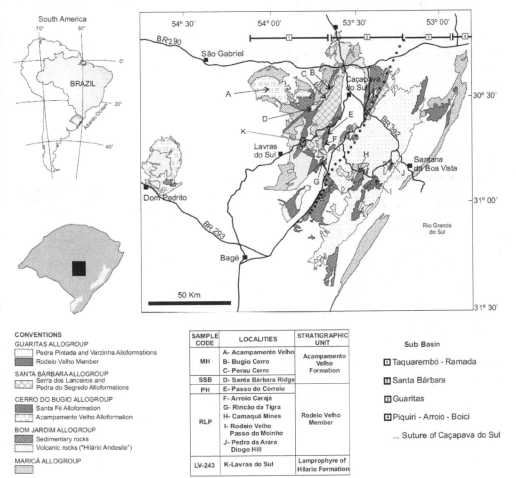

CONVENTIONS

GUARITAS ALLOGROUP
- Pedra Pintada and Varzinha Alloformations
- Rodeio Velho Member

SANTA BÁRBARA ALLOGROUP
- Serra dos Lanceiros and Pedra do Segredo Alloformations

CERRO DO BUGIO ALLOGROUP
- Santa Fé Alloformation
- Acampamento Velho Alloformation

BOM JARDIM ALLOGROUP
- Sedimentary rocks
- Volcanic rocks ("Hilário Andesite")

MARICÁ ALLOGROUP

SAMPLE CODE	LOCALITIES	STRATIGRAPHIC UNIT
MH	A- Acampamento Velho	Acampamento Velho Formation
	B- Bugio Cerro	
	C- Perau Cerro	
SSB	D- Santa Bárbara Ridge	
PH	E- Passo do Correio	
RLP	F- Arroio Carajá	Rodeio Velho Member
	G- Rincão da Tigra	
	H- Camaquã Mines	
	I- Rodeio Velho Passo do Moinho	
	J- Pedra da Arara Diogo Hill	
LV-243	K-Lavras do Sul	Lamprophyre of Hilario Formation

Sub Basin

1 Taquarembó - Ramada

2 Santa Bárbara

3 Guaritas

4 Piquiri - Arroio - Boici

... Suture of Caçapava do Sul

Fig. 1. Geological map showing the main Camaquã Basin stratigraphic units (modified from Paim et al, 2000) and the localization of the main regions.

According to Basei et al. (2000), the magmatic, metamorphic structural and geotectonic features in Southeastern South America recorded the superposition of two Neoproterozoic-Eopaleozoic orogenies: Brasiliano (700-620 Ma) and Rio Doce (620-530 Ma) orogenies. The end of the extensional regime related to the Brasiliano cycle is marked by late magmatism (600±10 Ma), conspicuous in Southern Brazil. The foreland basins, with sedimentation around 560±20 Ma and deformation around 530 Ma, represent the main volcanic-sedimentary record of Rio Doce Orogeny in Southern Brazil. The Camaquã Basin is considered to represent this type of basin. Differently from the current thoughts, Fragoso-Cesar et al. (2001) suggested an intraplate evolution to this basin, detached from the evolution of the Brasiliano Cycle. These authors related the development of the basin to the reactivation of the Brasiliano structures formed in the Neoproterozoic III to Eocambrian. They stated that the knowledge about the tectonic events and deformation of the sedimentary and volcanic rocks is still incipient, and there are no radiometric ages to position the deformation phases in time. They also ratify the connection between the Guaritas sedimentation and volcanics rocks from Rodeio Velho, and Paraná Basin. These units would be related to the development of SW-NE rifts during the initial stages of the evolution from Paraná Basin.

Robertson (1966)		Ribeiro et al. (1966)		Santos et al. (1978)		Leites et al. (1990)		Beckel (1990)		Paim et al. (2000)		
Camaquã Group	Coxilha Cong.	Grupo Camaquã	Coxilha Cong.				Volcano Sedimentary Sequence IV	Camaquã		Camaquã	Guaritas	Varzinha
												Pedra Pintada
	Guaritas Formation		Guaritas Formation						(6) Santa Bárbara Formation		Santa Bárbara	Pedra do Segredo
	(1) Santa Bárbara Formation		(2) Santa Bárbara Formation									Serra dos Lanceiros
											Cerro do Bugio	Santa Fé
	Ramada Rhyolite	Bom Jardim Group	Crespos Form. (3)	Maricá Group	Acampamento Velho Formation	Subgroup	Volcano-genic Sequence III		(3) Crespos Formation	Allosupergroup		Acampamento Velho
	Hilário Andesite		Arroio dos Nobres Form.(4)		(5) Cerro dos Martins Formation		Volcano Sedimentary Sequence II	Maricá Formation	(7) Arroio dos Nobres Form.	Bom Jardim		
	Maricá Formation		Maricá Formation		Pesse-gueiro Formation		Volcano Sedimentary Sequence I				Maricá	

Modified after Paim et al. 2000

Legend

— Unconformity type not especified
— Erosive unconformity
— Angular unconformity
- - - - Elongation surface
.... Transitional contact
← Stratigraphic limits usually used
∴ Bimodal volcanics
▲ Basic to intermediary volcanics
△ Acid volcanics

Notes: Includes, at the base, the Martins Andesite Member. At the base, contains Rodeio Velho Member. Composed by lower (Hilário) and upper (Acampamento Velho) members. Comprise the Vargas and Mangueirão members. Made up by Hilário and Arroio dos Nobres members. Constituted by the Rodeio Velho, at the base, Lanceiros, Varzinha and Guaritas members. Composed by Passo dos Bravos at the base, Vargas and Mangueirão members.

Table 1. Stratigraphic concepts for Camaquã Basin.

Silva et al. (2005) as cited in Borba et al. (2008) focusing on the tectonic evolution of Brasiliano events in the Mantiqueira Province (South and Central Brazil and Uruguay, Fig. 3) revealed three main orogenic systems: (a) Brasiliano I, with juvenile island-arc accretion and collisional climaxes at ca. 790 Ma (Embu Domain, Southeastern Brazil) and 730-700Ma (São Gabriel Orogeny, Southern Brazil); (b) Brasiliano II, with dominance of crustal

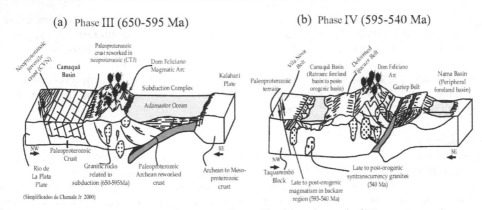

Fig. 2. Evolution model of the Sul-Riograndense Shield at the end of Neoproterozoic (Chemale Jr., 2000).

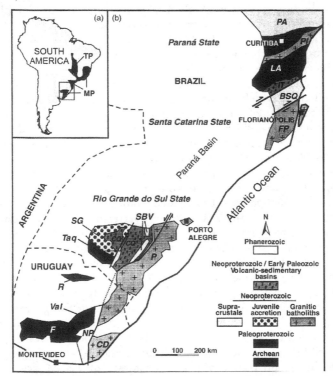

Fig. 3. Schematic geological map of the Southern portion of the Mantiqueira Province (b) located in Eastern South America (a); key to the terrains/blocks/belts/orogenies names: F= Florida; NP= Nico Perez; CD = Cuchilla Dionísio; Val= Valentines; R= Rivera; Taq=Taquarembó; SG= São Gabriel; SBV= Santana da BoaVista; P= Pelotas; FP= Florianópolis; BSQ = Brusque; LA= Luiz Alves; PI = Piên; PA= Paranapiacaba; unmetamorphosed volcanic-sedimentary successions are named as Camaquã (CQ) and Itajaí (IT) (Silva et al., 2005, in Borba et al., 2008).

recycling and collisional climaxes at 640-620Ma (Dom Feliciano Orogeny, Southern Brazil and Uruguay) and 600Ma (Paranapiacaba and Rio Piên Orogenies, Southeastern Brazil); and (c) Brasiliano III, also characterized by crustal recycling and collisional climaxes between 590-560Ma (Araçuai Orogeny) and 520-500Ma (Búzios Orogeny), recorded mainly in Eastern Brazil. Foreland basins associated with the Brasiliano II orogenic system comprise portions of volcanic-sedimentary (metamorphosed or unmetamorphosed) successions along the 3000 km of length from Mantiqueira Province.

2.2 Stratigraphy

Thick accumulations of sedimentary and volcanogenic rocks are present in the Camaquã and Itajaí basins of Southern Brazil, considered to be the "molassic fore deep of the Dom Feliciano Belt" by Fragoso-Cesar et al. (1984). These sedimentary successions evolved from marine to alluvial-aeolian continental strata and have been considered to be foreland basins along with their African counterparts (Gresse et al., 1996). Recent research considers the Camaquã Basin to be a series of the different basins (foreland, strike-slip and rift) with a shared locus of subsidence (Paim et al., 2000). In terms of stratigraphy, five major unconformity bounded units (Allogroups) have been defined by Paim et al. (2000): the Maricá, Bom Jardim, Cerro do Bugio, Santa Bárbara and Guaritas Allogroups (Table 1). They reflect the progressive continentalization of the Camaquã Basin. These units can be interpreted as the evolution of temporally and spatially restricted basins, evolved from a foreland retroarc into a rift (or hemigraben) tectonic setting. In this more recent interpretation, only the Maricá Formation was deposited in a foreland basin.

The Maricá Formation was defined by Leinz et al. (1941) as a basal unit of Camaquã Basin. This Formation is a ±2500 m-thick sedimentary package with coastal deposits, including fluvial to shallow marine sediments with scarce volcanic contribution (Leites et al., 1990). The nature of the basal contact of this unit with the Neoproterozoic juvenile terrain is difficult to be defined, either an unconformity or a thrust plane, due to low relief, thick soil profiles and vegetation, and widespread younger intrusive bodies. According to Borba (2004), the Maricá formation can be divided into three major packages: the lower, intermediate and upper successions. The lower succession comprises conglomeratic and coarse-grained sandstones with very well rounded granite–gneiss pebbles and cobbles either marking the stratification planes or dispersed through the beds. The intermediate succession is made up of indurated siltstones and shales, either massive or showing plane-parallel lamination, wave and climbing ripples, and lenticular lamination, as well as hummocky sandstones, suggesting a marine origin. The upper succession is very similar to the lower one, with the sandstones featuring large-scale troughcross and low-angle stratifications. Well-rounded granite–gneiss and rare volcanic, rhyolitic pebbles and cobbles are also present within the sandstone beds. The lower and upper successions correspond to braided fluvial deposits, while the intermediate package is related to a shallow marine depositional environment. The strata of the Maricá Formation were strongly affected by subsequent events of volcanism, intrusion, sedimentation and uplift. Volcanic rocks of the Bom Jardim Group (andesite Hilário) and the Acampamento Velho Formation (rhyolitic flow, rhyolitic pyroclastic rocks and basic flow) caused metamorphism contact in the Maricá Formation. The unit is unconformably overlain by the alluvial, fluvial and eolian deposits of the Santa Bárbara and Guaritas formations (Borba & Mizusaki, 2003). A syn-depositional volcanic event (Santos et al., 1978) has been recognized in the rock record of the Maricá Formation, evidenced by the occurrence of volcanogenic beds interlayred with sedimentary rocks of the lower succession. Circular structures resembling degassing pipes also suggest the activity of volcanic hot springs (Borba et al., 2004). According to Paim et al (2000), in terms of deformation at the Taquarembó/Ramada sub-basin, are recognizes reverse ruptile faults and weak folds, while the Piquiri/Boici sub-basin is dominated by structures associated to the transcurrent processes in ruptile to ductile-ruptile structural level.

The Bom Jardim Group (Paim et al, 2000), is a volcano-sedimentary sequence that shows a lithostratigraphic complexity in its area-type and similarity lithological of some its units with others from Camaquã Supergroup units (Table 1). Bom Jardim Group, in its area-type (Janikian et al., 2003), shows approximately 4,000 thicknesses and constituted of andesitic volcanic rocks, pyroclastic rocks (lapilli tuffs, lithic tuffs and thick vitreous tuffs – Hilário Formation) and sedimentary rocks. These rocks were generated in an active tectonic lacustrine basin. Robertson (1966) made the first attempted correlation among the volcano-sedimentary units from Bom Jardim area and others expositions from Camaquã Supergroup. Ribeiro et al. (1966) and Tessari & Picada (1966) correlated the outcropping units from Bom Jardim and Vale Piquiri (Piquiri Sub-basin) areas, covering the Arroio dos Nobres Formation.

The Hilário term was first used by Robertson (1966) to designate a variety of rock types composed of mainly andesitic to dacitic composition (flows, tuffs, volcanic conglomerates, mudflows, dykes and intrusive mafic rocks). Ribeiro et al. (1966) formalized this unit as a member of Crespos Formation - Bom Jardim Group. Santos et al. (1978) used the Hilário term to refer a member of Cerro dos Martins Formation, while Horbach et al. (1986) upgraded these rocks to the formation level. Several studies have been carried out Hilário Formation rocks. Nardi & Lima (1985), Porcher et al. (1995), Lima & Nardi (1998) works are especially relevant. Thus, Lima & Nardi (1998) defined the Lavras do Sul Shoshonitic Association, which comprises a wide compositional range of intrusive and extrusive basic to acid rocks. The extrusive rock includes the Hilário Formation and the spessartitic lamprophyres. In the Bom Jardim Group, the volcanic rocks are mostly in the lower stratigraphic levels, as lava flows or small shallow intrusion.

It the Santa Bárbara Sub-basin (Paim et al., 2000) toward the Guaritas Sub-basin, the volcanic rocks of this group are replaced gradational by alluvial conglomerates rich in volcanic clasts from basic to intermediate composition, and more distally, by sandy and silty related to the bottom flows (turbidites) , which are bounded to underwater deltaic systems. Next to the high topographic that today subdivides the Santa Bárbara and Guaritas sub-basins (e.g. Alto Caçapava), is common the occurrence of the erratic conglomerates containing local derivation clasts (shales and marble), indicating that the partitioning of the Camaquã Basin in several sub-basins occurred prior to the deposition of this unit.

Cerro do Bugio Allogroup (or Group) is composed of volcanic rocks at the base (Acampamento Velho Alloformation or Formation) and alluvial conglomerate deposits from Santa Fé Alloformation (ou Formation). According Paim (2000), Cerro do Bugio is limited by two angular unconformity that delineate its contact with the lower (Bom Jardim Allogroup) and upper (Santa Bárbara Allogroup) units. Cerro do Bugio Unit shows approximately 500 m thick and is composed of, from the base to the top, Acampamento Velho (volcanic) and Santa Fé (alluvial conglomerates, and secondarily, by sand-pelitic rhytmits and pelites), separated by an erosive unconformity. The Santa Fé Formation consists mainly of alluvial conglomerates rich in volcanic and plutonic clasts of the acid composition, which vary vertically to sand-pelitic rhytmits. These lithologies represent the alluvial systems interlaced, considering the transversal (deltaic fans, conglomeratic with paleocurrents for SE) and longitudinal (fans of the interlaced plain) characters. This allogroup was affected by dip-slip and transcurrent faults generated in crustal to ductile-ruptile level. Considering the field work observations, the Acampamento Velho Formation volcanic rocks are often interdigitated with rocks from Santa Fé Formation, suggesting the contemporaneity between both units.

Acampamento Velho Formation (sensu Cordani et al., 1974) is the basal unit of the Santa Bárbara Group and crops out mainly in the Santa Bárbara and Taquarembó-Ramada sub-basins. The basal contact of the Acampamento Velho Formation is unconformable with the Maricá and Bom Jardim Groups. According to Almeida et al. (2002), in the Cerro do Bugio , Perau and Serra de Santa Bárbara (Fig. 1), the volcanic rocks (basaltic and rhyolitic compositions) upper contact is a disconformity by the alluvial conglomerates, and locally, by deltaic-lacustrine sandstones and mudstones from Santa Fé Formation.

The Santa Bárbara Group, defined by Robertson (1966), Ribeiro et al., (1966) and Ribeiro & Fantinel (1978) as Santa Bárbara Formation, comprises red-colored, sandstones and siltstones, and includes the Acampamento Velho and Santa Bárbara formations. It was named as Santa Bárbara Allogroup by Paim et al. (2000). This group is composed of Acampamento Velho, Santa Fé, Lanceiros and Segredo Formations, from the base to the top (Table 1). According to the same authors, the Lanceiros Formation is related to the progradation of an interlaced and sandy deltaic system of the longitudinal character, and the Pedra do Segredo Formation can be bounded to the progradation of an interlaced and sandy-conglomeratic of transversal nature. Borba & Mizusaki (2003) organized the Santa Bárbara Formation rocks in three depositional sequences named as, from the base to the top: Sequence I, Sequence II and Sequence III. The two basal sequences (I and II) represent a coherent depositional pattern, with axial fluvial and deltaic fan systems, which deposit North-Eastward with lateral contribution from alluvial fans. Such coarse deposits are composed mainly of metamorphic clasts derived from Alto Caçapava and Eastern steep Santa Bárbara Basin. The Sequence I shows a dip ranging from 32º to 40º NE in the Northern outcrops and 10º to 16º NE in the South, and the Sequence II from 28º to 30º SE. The Sequence III lies unconformable over basal subunits and reflects the axial system inversion, in that the gravel bed deposits paleocurrents systematically point South/South-Westward. The alluvial fan deposits of the Sequence III also suggest a tectonic rearrangement basin, with partial erosion from basal sequences, and the presence of granitic composition fragments, which reflects the deeper denudation stage of the Alto Caçapava and possibly a significant hiatus at the Sequence III base. The third dip sequence is between 23º to 26º ENE. The Acampamento Velho Formation is not included in the Santa Bárbara Formation (Borba & Misuzaki, 2003).

The Guaritas was defined by Goñi et al. (1962), Ribeiro et al (1966), Robertson (1966) and Ribeiro & Fantinel (1978) and is lithologically similar to the Santa Bárbara Group. This unit represents the last great depositional event preserved in Camaquã Basin and is above unconformity of the previous units. It was denominated Guaritas Allogroup (Paim et al., 2000) and is composed of two alloformations or formations, from the base to the top: Pedra Pintada and Varzinha (Table 1). These units are bounded by an erosive unconformity. The Guaritas Group shows around 800 m thick and represents the last depositional episode preserved within of the Camaquã Basin. The Pedra Pintada Formation is temporal and spatially related to the volcanic rocks named as Rodeio Velho Member. It was originated in a desert environment with crescent eolian dunes, in an inter-dune area that describes interchange of dry and wet periods, and basal level plains that comprehend dry seasons during wetter periods (Paim et al., 2000). It is characterized by fine to medium sandstone, well selected, with large to very-large size cross-stratification and bulk sandstones, and occurs secondarily mudstones and fine to medium sandstones with crossed lamination by stream and waves (Paim et al., 2000). The rocks that belong to this formation are found in the Santa Bárbara and Guaritas sub-basins (Fig. 1). The Varzinha Alloformation is composed of sedimentary faces that represent a braided fluvial system at the West Guaritas Sub-basin portion, and a fanlike alluvial system at the East portion. According to Paim et al. (2000), analyzing the paleocurrents data of this unit, we detect a lateral association of two different alluvial systems: (1) interlaced fluvial system, West side of the Guaritas sub-basin; (2) alluvial fans represented by, at the least two lobes, at the East side of the Guaritas Sub-basin (Paim, 1995). The Varzinha Alloformation upper levels correspond to the several sets of the progradacional parasequences from deltaic source. They were developed de within a shallow lacustrine basin. The deltaic deposits were associated to a tributary deltaic fan system (lateral contribution) at the East edge of the Guaritas Sub-basin, and an interlaced plain delta system (longitudinal contribution), which shows progradation to the SW, at the Eest edge of the Guaritas Sub-basin. Almeida et al. (2003a) described the Varzinha Alloformation as a covering of the area that present four cones (Rodeio Velho Member) at Santa Bárbara Sub-basin (see iten 5). According to Takehara et al (2010), the Guaritas Group, which comprises the aeolian and alluvial plain as well as fan deposits, is covering the Santa Bárbara Group rocks in Santa Bárbara Sub-Basin.

Rodeio Velho Member (sensu Ribeiro et al., 1966) has been described as being at least three vesicular andesite flows, with thickness estimated at 100 m and no evidence of explosive activity. Silva Filho (1996) showed the intrusive character of the magmatism, rejecting to form ideas of an exclusively volcanic event. Fragoso Cesar et al. (2000) named these rocks as Rodeio Velho Intrusive Suite, which is represented by tabular intrusions within the sub-horizontal continental deposits of the Guaritas Group. Almeida et al. (2000, 2003a) stated that the before mentioned event has a basaltic andesitic composition with alkaline affinity and the rocks cropped out as lava flows, pyroclastic deposits and shallow intrusions.

3. Geochronological data

We present a synthesis of geochronological ages for the different lithologies mentioned in the literature as well as new U-Pb ages obtained by us in samples from Maricá Formation (sandstone), Hilário Formation (lamprophyre), Acampamento Velho base (andesitic basalt) and Rodeio Velho Member (alkaline basalt).

3.1 Analytical procedures used for new U-Pb ages

All zircons were mounted in epoxy with 2.5 cm diameter and polished until the zircons were just revealed. Images of zircons were obtained using the optical microscope (Leica MZ 12$_5$) and back-scatter electron microscope (Jeol JSM 5800). Zircon grains were dated with laser ablation microprobe (New Wave UP213) coupled to a MC-ICP-MS (Neptune) at the Isotope Geology Laboratory from UFRGS. Isotope data were acquired using static mode with spot size of 25 and 40 um. Laser-induced elemental fractional and instrumental mass discrimination were corrected by the reference zircon (GJ-1) (Simon et al., 2004), following the measurement of two GJ-1 analyses to every ten sample zircon spots. The external error is calculated after propagation error of the GJ-1 mean and the individual sample zircon (or spot).

Laser operating conditos	
Laser type New Wave UP213	**MC-ICP-MS Neptune**
• Laser output power 6 J/cm^2 • Shot repetition rate 10 Hz • Laser spot 25 and 40 μm	• Cup configuration: Faradays ^{206}Pb, ^{208}Pb, ^{232}Th, ^{238}U MIC's ^{202}Hg, ^{204}Hg + ^{204}Pb, ^{207}Pb • Gas input: Coolant flow (Ar) 15 l/min Auxiliary flow (Ar) 0.8 l/min Carrier flow 0.75 l/min (Ar) + 0.45 l/min (He) Acquisition 50 cycles of 1.048 s

The common ^{204}Pb, after Hg correction based on ^{202}Hg simultaneously measured, is insignificant in most situations. For instance, typical signal intensity due to ^{204}Hg during a laser ablation on standard Zircon is 600-1000 cps range, while the calculated count rate for ^{204}Pb is less than statistical error of ca. 25-33 cps. We assume that the ^{204}Pb values obtained from zircons have common Pb composition, assuming concordant age of ^{206}Pb/^{238}Pb and ^{207}Pb/^{206}Pb (as estimated age). In this case, we estimate the radiogenic composition of ^{206}Pb and Pb^{207} using the equation as fraction of non radiogenic ^{206}Pb (Williams, 1998):

$f_{206} = [^{206}Pb/^{204}Pb]_c / [^{206}Pb/^{204}Pb]_s$

$f_{207} = [^{207}Pb/^{204}Pb]_c / [^{207}/^{204}]_s$

For common lead isotope composition, we assumed an isotope composition evolution proposed by Stacey and Kramers (1975), which is required to attribute an initial estimated age.

3.2 Geochronological data for different units

Soliani Jr. et al. (1984) presented a synthesis of all the radio-chronological determinations obtained for the crystalline and sedimentary rocks from Sul-Riograndense Shield meridional portion, including part of the Santa Catarina, the whole Rio Grande do Sul and Uruguay, thus preserving the Cordani et al (1974) and Teixeira (1982) proposal, which says that the West Sul-Riograndense Shield region would have had an Transamazonic evolution, and the meridional portion would have been generated in the Brasiliano Cycle. Sartori & Kawashita (1985) obtained the age of 550 Ma for Caçapava do Sul Granitic Batholith using 21 Rb-Sr isotope analyses on total-rock.

According to Soliani et al. (2000), the first work of the regional integration and geochronological synthesis was made by Cordani et al (1974). The authors relate the K-Ar and Rb-Sr radiometric data to the Eopaleozoic rocks, which fill the Camaquã Basin. Thus, the Rb-Sr data of the Eastern portion with syntectonic characteristics (granites and migmatites) and post-tectonic (isotropic granites) show respectively, values from 650-600 Ma. K-Ar ages of the same region are in the same coherent interval characteristics of the Brasiliano Cycle. For the Midwest region of the shield were found radiometric values that suggest an older age, although a significant data percentage show Brasiliano ages for the granitic massive such as the Caçapava do Sul and Lavras do Sul among others. In the same way was confirmed that the volcanic rocks (Hilário and Acampamento Velho formations) were generated at the end of the Pre-Cambrian or at the beginning of the Phanerozoic, exactly as has been commented by Minioli & Kawashite (1971).

Leite et al. (1995, 1998) and Leite (1997) mentioned for the Rio Grande do Sul region, the first U-Pb data obtained in zircons using the sensitive high mass-resolution ion microprobe method, being the isotope data for the Caçapava do Sul granitic massives of the 561± 6 Ma and 541± 11 Ma. The first age is considered as corresponding to the igneous protolith that generated the Caçapava Granite. The second age is related to a magmatic event that generated the granite. Remus et al. (1996, 1997) used sensitive high mass-resolution ion microprobe method to obtain the U-Pb ages in zircons for the Lavras do Sul Granite core and edge, respectively: 592±5 and 580 ±5 Ma. The age of 565± 14 Ma was obtained for the Caçapava do Sul Granite.

3.2.1 Maricá Formation

The first geochronological investigation in Maricá Formation was performed by Borba et al. (2008). These authors made analysis of zircons from pyroclastic cobbles, which yielded an age of 630.2±3.4 Ma (2σ), interpreted as the age of syn-sedimentary volcanism, and thus of the deposition itself. This result indicates, according to Borba et al. (2008), that Maricá Formation was deposited during the main collisional phase (640–620 Ma) of the Brasiliano II Orogenic system from Silva classification (2005). Paim et al (2000) suggested a depositional age between 620 and 600 Ma in a foreland basin.

Spot number	Isotope ratios [2,3] 207Pb/235U	1s	206Pb/238U	1s	Rho[4]	207Pb/206Pb	1s	Age (Ma) 206Pb/238U	±	207Pb/235U	±	207Pb/206Pb	±	232Th/238U[1]	%of Conc[5]	Apparent age	±
156-A-I-01	0,88525	3,2	0,1064	2,3	0,73	0,0604	2,2	652	15	644	20	616	13	1,18	106	616	13
156-A-I-02	0,79543	6,5	0,0965	4,8	0,74	0,0598	4,3	594	28	594	38	596	26	0,67	100	596	26
156-A-I-07	0,81604	2,4	0,0988	1,4	0,59	0,0599	2	607	9	606	15	600	12	1,49	101	600	12
156-A-I-10	0,85889	2,7	0,1028	1,5	0,55	0,0606	2,2	631	9	630	17	626	14	1,08	101	626	14
156-A-I-11	0,78707	3,7	0,0949	2,4	0,65	0,0601	2,8	585	14	590	22	608	17	1,02	96	608	17
156-A-I-14	0,76104	4,1	0,0904	3,9	0,96	0,0611	1,1	558	22	575	23	642	7	0,11	87	642	7
156-A-I-25	0,77688	3,4	0,0925	2,3	0,69	0,0609	2,4	570	13	584	20	636	16	0,60	90	636	16
156-A-I-37	0,84301	2,8	0,1007	1,6	0,56	0,0607	2,3	619	10	621	17	629	15	0,86	98	629	15
156-A-I-09	4,67403	2,02	0,31130	0,74	0,37	0,1089	1,9	1747	25	1763	42	1781	34	0,61	98	1781	34
156-A-I-03	5,9604	2,5	0,3495	2,2	0,88	0,1237	1,2	1932	43	1970	50	2010	24	0,38	96	2010	24
156-A-I-18	6,07717	2,5	0,3555	1,9	0,74	0,124	1,7	1961	37	1987	50	2014	34	1,54	97	2014	34
156-A-I-19	6,47889	1,4	0,3709	1,1	0,75	0,1267	0,9	2033	22	2043	29	2053	19	0,08	99	2053	19
156-A-I-28	10,3139	2,5	0,4624	1,7	0,71	0,1618	1,7	2450	42	2463	60	2474	43	0,90	99	2474	43
156-A-I-33	9,85995	1,5	0,4457	1,1	0,72	0,1604	1	2376	26	2422	36	2460	25	0,32	97	2460	25
156-A-I-36	10,2057	2,6	0,4624	1	0,37	0,1601	2,4	2450	24	2454	63	2457	59	0,53	100	2457	59
156-A-I-13	9,68137	1,4	0,4187	0,9	0,62	0,1677	1,1	2255	19	2405	33	2535	27	0,42	89	2535	37
156-A-I-06	11,4312	1,9	0,4747	1,5	0,77	0,1747	1,2	2504	37	2559	49	2603	32	0,73	96	2603	32
156-A-I-23	11,371	2,3	0,4715	0,9	0,4	0,1749	2,1	2490	23	2554	58	2605	54	0,43	96	2605	54
156-A-I-27	11,4641	2,6	0,4738	1,9	0,71	0,1755	1,9	2500	47	2562	68	2611	49	0,73	96	2610	49

Table 2. U/Pb zircon data of the sandstone sample (CK-239D) for Maricá Formation obtained by in situ LAM-ICPMS-MC.

One sandstone sample (CK-239-D) used in the U-Pb analysis was collected at North of the Estância 3 Estradas (UTM 771863- 6576404) in Coxilha do Tabuleiro Sheet (1:50.000). Twenty zircon crystals from sample CK-239-D were dated. Based on the dating results, we detected the presence of follow age set: 2606 Ma (4 zircons), 2473 Ma (3 zircons), 2050 Ma (3 zircons) and 1790 Ma (one zircon - Table 2, Fig. 4). The 601 ±13 Ma age (based on eight zircons – Table 2, Fig. 4 - Early Ediacaran) was considered as the age older from Maricá Formation (Fig. 6). All the zircons that showed older ages were considered xenocrystal zircons. Thus, the magma would have assimilated Neoarchean (NA), Paleoproterozoic (Siderian, Rhyacian and Statherian) material during its ascension. The selected zircons that allowed the Maricá age setting (Fig. 5), in general are colorless and dirty, all grains are between 175 μm and 344 μm in size and show diffuse zonation and subrounded contours. The zircons do not have impurities and inclusions, but show few fractures. In the A-I-02 zircon (Fig. 5) is observed an incomplete concentric fracturing at the edge of the crystal.

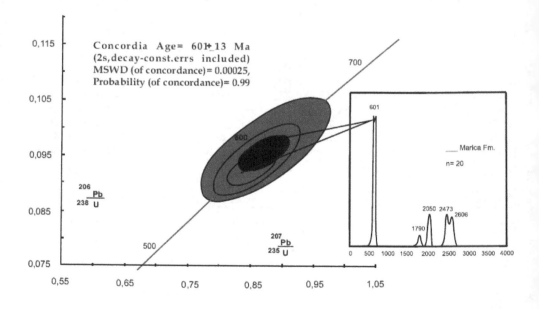

Fig. 4. U-Pb concordia data diagram with tracing considering only the younger zircon ages from Maricá Formation sandstone sample. a: U-Pb age histogram considering all the obtained ages (table 2). We detected the presence of five ages: 2606 Ma, 2473 Ma, 2050 Ma, 1790 Ma and 601 Ma. The 601 Ma age is considered as the older from Maricá Formation.

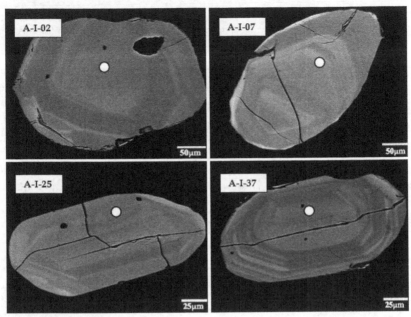

Fig. 5. Zircons microphotographies that defined the 601±13 Ma age from CK239D sample for Maricá Formation sandstone. A-I-02 = sample number, o = spot showing the place where the analyses were obtained in each zircon. Data showed in table 2.

3.2.2 Hilário Formation

Ribeiro & Teixeira (1970) mentioned values between 510-535 Ma for Hilário Andesite. These values were obtained through the K-Ar method. Cordani et al. (1974) obtained ages of 510 and 535 Ma using the Rb/Sr method. $^{206}Pb206/^{238}U$ data (sensitive high mass resolution ion microprobe) were obtained for Hilário Andesite by Remus et al. (1999) and gave an age of 580 Ma. Janikian et al (2008) mentioned ages from 590 to 585 Ma, which are interpreted for us as representative of Hilário Formation extrusion, being U-Pb age of 590 ± 5.7 Ma obtained in the Hilário Andesite as the most representative. The ages obtained by Remus et al. (2000) using the U-Pb (sensitive high mass-resolution ion microprobe) for Lavras Granite (592±5 Ma, 597±5 Ma, 580±7 Ma) and Caçapava do Sul (589±5 Ma) are considered as comagmatic to the Bom Jardim volcanism. Chemale Jr. et al. (2000) reported an age of 592 Ma for Andesite Hilário.

The LV- 243 sample (alkaline basalt) dated using the U-Pb method by us was collected 15 km away from the Lavras do Sul, following by RS 357 road toward Caçapava do Sul (Fig.1). Lima & Nardi (1998) said that the out crop is a dome with size around 0.5 km² and a circular shape. It shows a marked columnar disjunction with hexagons of variable diameter, but no exceeding 15 cm diameter. Nine zircon crystals (sample LV- 243) were dated (Table 3). The obtained age of 591.8 ±3.0 Ma was calculated based on individual zircon ages from Table 3. This age defines the lamprophyre age (Fig. 6), which belong to the Hilário volcanism manifestation. The selected zircons that allowed the lamprophyres ages setting, in general, are colorless and clean, all grains are between 150 μm and 333 μm in size and the most show very clear zoning. The grains also show prismatic and pyramidal shapes, incomplete and partially serrulated contours and few fractures (Fig. 7).

Spot number	Concordia 1					Age (Ma)								
	$^{207}Pb/$ ^{235}U	error	$^{206}Pb/$ ^{238}U	error	Rho 1	^{206}Pb ^{238}U	error	^{207}Pb / $_{235}U$	error	^{207}Pb / $_{206}Pb$	error	^{232}Th / $_{238}U$	% Disc	f206
072-B-II-1	0,80351	3,22	0,09732	1,80	0,56	599	11	599	19	599	16	0,34	0	0,0010
072-B-II-2a	0,80971	3,22	0,09779	1,47	0,46	601	9	602	19	605	17	0,53	1	0,0005
072-B-II-2b	0,81989	3,00	0,09891	1,46	0,49	608	9	608	18	608	16	0,33	0	0,0009
072-B-II-3	0,80923	3,56	0,09799	1,58	0,44	603	10	602	21	600	19	0,34	0	0,0012
072-B-II-4	0,78691	3,08	0,09534	1,37	0,45	587	8	589	18	599	17	0,34	2	0,0004
072-B-II-7	0,78269	3,54	0,10586	1,80	0,51	649	12	638	23	602	18	0,41	-8	0,0004
072-B-II-9	0,76140	3,95	0,09280	1,59	0,40	572	9	575	23	586	21	0,41	2	0,0003
072-B-II-16a	0,74234	4,22	0,08980	3,22	0,76	554	18	564	24	602	16	0,45	8	0,0002
072-B-II-16b	0,79361	3,39	0,09618	1,57	0,46	592	9	593	20	598	18	0,34	1	0,0007

Table 3. U/Pb zircon data of the lamprophyre sample (LV-243) for Hilário Formation obtained by in situ LAM-ICPMS-MC.

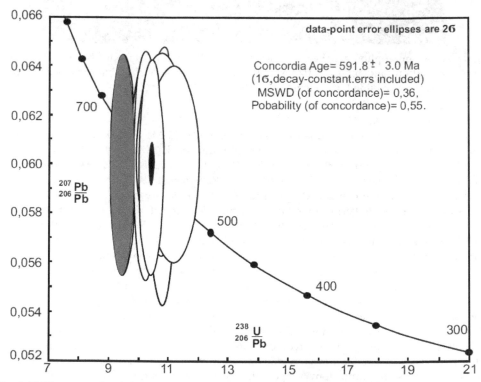

Fig. 6. U-Pb concordia data diagram with tracing considering the zircon ages for Hilário Formation lamprophyre (LV-243 sample). Gray ellipse cores represent the single analysis from zircons of the table 3. Black ellipse represents the concordia age for LV-243 sample, calculated from single zircon ages of the table 3.

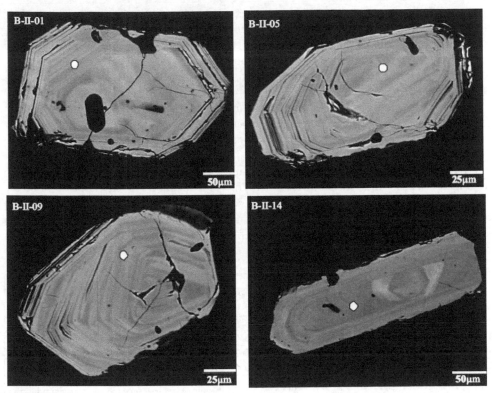

Fig. 7. Zircons microphotographies that defined the 553 ± 5.4 Ma age for lamprophyric rock
from Hilário Formation (LV-243 sample). A-I-02 = sample number, o = spot showing the
place where the analyses were realized in each zircon. Data showed in table 3.

3.2.3 Acampamento Velho Formation

The first geochronological investigation of the rhyolites was performed by Cordani et al.
(1974) followed by Sartori (1978) and Teixeira (1982). Soliani Jr. (2000) compiled their data
and obtained an age of 529 ± 4 Ma and R_0 = 0,706 (Rb-Sr whole rock). Another Rb-Sr dating
was performed by Almeida et al. (2002), who studied the rhyolitic flows from Cerro do
Bugio area and the dykes intruding the Maricá Formation. These authors obtained two
whole rock isochrons: 545.1 ± 12.7 Ma (R_0 = 0.709) and 546 ± 12.9 Ma (R_0 = 0.714). However,
Sommer et al. (2005) presented U-Pb data (sensitive high mass-resolution ion microprobe)
using eleven zircon crystals from rhyolites of Vila Nova do Sul area (Ramada Plateau),
obtaining an age of 549.3 ± 5 Ma, which is more reliable. Therefore, all the obtained ages
indicate that the rocks from Acampamento Velho Formation upper felsic association belong
to the Early Neoproterozoic III (NP_3).

The MH-13 sample (andesitic basalt) was selected to obtain the absolute age from lower
mafic association of Acampamento Velho Formation base. Nineteen zircons crystals were
dated (Table 4). Based on the dating results, we detected the presence of four age group:
2799 ± 21 Ma (one zircon); 2182 ± 55 Ma (one zircon), 612.6 ± 15 Ma (seven zircons) apart

from anchored age, that corresponds to intersection at 2442 ± 54 Ma (three zircons) (Table 4, Fig. 8a). The 553 ± 5.4 Ma age (based on six zircons – Table 4, Fig. 8b) was considered as the lower mafic association age. All the zircons that showed older ages were considered xenocrystal zircons. Thus, the magma would have assimilated Neoarchean (NA), Paleoproterozoic (Siderian and Rhyacian) and older Neoproterozoic material during its ascension.

The selected zircons that allowed the Acampamento Velho Formation base age setting, in general are colorless and dirty, all grains are between 87 μm and 158 μm in size and show diffuse zonation, prismatic and pyramidal shapes, incomplete and partially serrulated contours, few fractures and inclusions (Fig. 9). The fain and broad zoning suggest that it may have been formed by very slow and complex crystallization of a magma body with prolonged residence time in the lower crust. The crystallization velocity appears to be the major controlling factor of the elongation ratio to the zircon. Skeletal zircon crystals are the most extreme form of rapid growth. It is relevant to point out that the Acampamento Velho Formation mafic flow dated zircons show morphologic characteristics of a slow generation in the magmatic chamber.

Spot number	Concordia 1					Age (Ma)							%	
	$^{207}Pb/$ ^{235}U		$^{206}Pb/$ ^{238}U		Rho1	$^{206}Pb/$ ^{238}U		$^{207}Pb/$ ^{235}U		$^{207}Pb/$ ^{206}Pb		$^{232}Th/$ ^{238}U	Dis.	F206
119-A-I-02	0,73099	4,86	0,09051	2,05	0,42	559	11	557	27	551	24	0,62	-1	0,0007
119-A-I-12	0,71742	5,44	0,08906	2,96	0,54	550	16	549	30	546	25	0,51	-1	0,0008
119-A-I-24	0,71164	4,02	0,09119	3,43	0,85	563	19	546	22	476	10	0,69	-18	0,0000
119-A-I-28	0,71175	2,90	0,08921	1,89	0,65	551	10	546	16	525	12	0,99	-5	0,0000
119-A-I-32	0,72048	4,18	0,08986	2,12	0,51	555	12	551	23	535	19	0,66	-4	0,0000
119-A-I-26	0,68233	4,88	0,08835	4,35	0,89	546	24	528	26	453	10	1,06	-21	0,0000
119-A-I-05	0,77464	3,54	0,09524	1,96	0,55	586	11	582	21	567	17	0,52	-3	0,0003
119-A-I-07	0,83567	4,54	0,10066	2,19	0,48	618	14	617	28	611	24	1,32	-1	0,0005
119-A-I-16	0,79403	4,41	0,09638	1,88	0,43	593	11	593	26	595	24	0,96	0	0,0009
119-A-I-19	0,81080	3,32	0,09729	2,50	0,75	599	15	603	20	619	14	0,50	3	0,0000
119-A-I-22	0,84916	5,93	0,10603	2,60	0,44	650	17	624	37	533	28	0,49	-22	0,0001
119-A-I-23	0,83998	3,39	0,10432	1,63	0,48	640	10	619	21	545	16	0,65	-17	0,0000
119-A-I-29	0,77999	4,20	0,09460	3,70	0,88	583	22	585	25	596	12	0,66	2	0,0002
119-A-I-20	14,70031	2,19	0,54875	1,28	0,58	2820	36	2796	61	2779	49	0,72	-1	0,0001
119-A-I-01	10,30577	2,14	0,46719	1,94	0,90	2471	48	2463	53	2456	22	0,23	-1	0,0011
119-A-I-10	10,20802	2,51	0,46490	1,98	0,79	2461	49	2454	62	2448	38	0,34	-1	0,0005
119-A-I-18	9,37569	1,89	0,43936	0,99	0,52	2348	23	2375	45	2399	39	1,07	2	0,0018
119-A-I-17	4,33522	2,14	0,28940	1,58	0,74	1639	26	1700	36	1777	26	0,73	8	0,0010
119-A-I-01	7,59979	6,12	0,40841	4,51	0,74	2820	36	2185	134	2163	89	0,41	-2	0,0017

Table 4. U/Pb zircon data of the andesitic basalt sample (MH-13) for Acampamento Velho Formation base obtained by in situ LAM-ICPMS-MC.

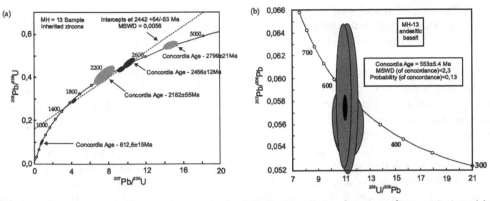

Fig. 8. Concordia diagram from Acampamento Velho Formation lower mafic association: (a) U-Pb data diagram with tracing considering only the older zircon ages of the 553 Ma (table 4). We detected the presence of four ages: 2799 ± 21 Ma, 612.6 ± 15 Ma, 612.6 ± 15 Ma and 2442 ± 54 Ma. (b) U-Pb data diagram with tracing considering only the younger zircon ages from Acampamento Velho Formation lower mafic association (MH-13 sample). Gray ellipse cores represent the single zircon analysis (table 4). Black ellipse represents the concordia age of the MH-13 sample calculated from single zircon ages (table 4).

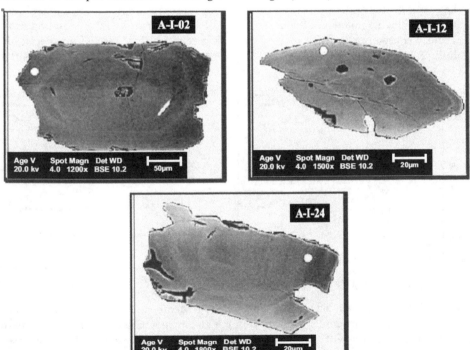

Fig. 9. Zircons microphotographies that defined the 553 ± 5.4 Ma age for Acampamento Velho Formation base. A-I-02 sample number, O = spot showing the place where the analysis were realized in each zircon. Data showed in table 4.

3.2.4 Rodeio Velho Member

Hartmann et al. (1998) determined the U-Pb age of 470 ± 19 Ma (sensitive high mass-resolution ion microprobe method) for this event (Middle-Ordovician), which suggest that Rodeio Velho Member is related to the beginning oh the Paraná Basin formation. On the other hand, Chemale Jr. (2000) obtained Sm-Nd model ages (T_{DM}) from 1.6 to 1.9 Ga, suggesting a modified mantle origin for these rocks. Rodeio Velho Member was probably the source of heat and hydrothermal solutions for Cu (Ag, Au), Pb and Zn (Cu, Ag) mineralizations in the Camaquã Mines (Lima et al., 2001). Hydrothermal illites collected in those mines also yielded K-Ar ages around 465 Ma (Bonhomme & Ribeiro, 1983).

We used the sample RLP-20 (alkaline basalt) to date Rodeio Velho event using U-Pb method. The sample was collected between Rodeio Velho and Pedra de Arara area (Fig. 1). At this area, the volcanic rocks show two lava flows, being the lower flow more vesiculated (flow top) than the upper, which is massive. This is related to the sandstones with great cross-stratification, suggesting the eolian character of this rock and it would correspond to the Pedra Pintada Formation from Guaritas Group. We dated sixteen zircon crystals from sample RLP-20. Based on these dating results, we detected the presence of 4 age groups: 2190± 18 Ma (one zircon); 1079 ± 12 Ma (six zircons); 658 ± 8.3 Ma (two zircons) and 547 ± 6.3 Ma (based on five zircons), (Table 5, figs. 10a and 10b). The younger obtained age, that was calculated based on individual zircon ages from Table 5, defined the Rodeio Velho Member age (Fig. 10a). All the zircons that showed older ages were considered xenocrystal inherited zircons. Thus, the magma would have assimilated Paleoproterozoic (PP₂), Mesoproterozoic (MP₃) and Neoproterozoic (NP₂) rocks during its ascension. The selected zircons used to obtain the Rodeio Velho Member age are in general (Fig. 11) colorless and transparent, the

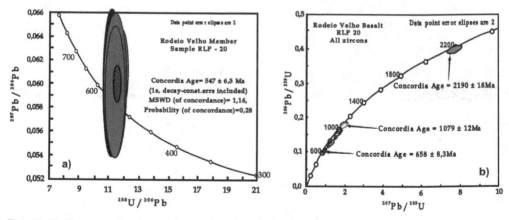

Fig. 10. U-Pb concordia data diagram for Rodeio Velho Member. (a) - U-Pb concordia data diagram with tracing considering only the younger zircon ages. Gray ellipse cores represent the single zircon analysis, with exception of A-I-02 zircon (black ellipse). Dark gray ellipse represents the concordia age calculated from single zircon ages (Table 5). The A-I-02 zircon sample shows high percentage of discordance and was not included in the calculating data (black ellipse). (b) - U-Pb concordia data diagram with tracing considering only the older zircon ages of the 547 Ma (table 5).

size varies between 202 µm and 321 µm, the crystals show zonation, euhedral to subhedral shape, contours partially serrulated, few fractures, impurities and inclusions. They are very small and occupy the crystal rims. The A-I-19 zircon (Fig. 11d) is constituted by an older core with age between 983 Ma and 1000 Ma (spot a, figure 11d) and 545 to 554 Ma rims (spot b, figure 11d).

Spot number	Concordia 1					Age (Ma)						$^{232}Th/$ ^{238}U	%	
	$^{207}Pb/$ ^{235}U		$^{206}Pb/$ ^{238}U		Rho 1	$^{206}Pb/$ ^{238}U		$^{207}Pb/$ ^{235}U		$^{207}Pb/$ ^{206}Pb			Dis	F206
072-A-I-02	0,72563	3,62	0,08724	2,32	0,64	539	12	554	20	615	17	1,60	12	0,0010
072-A-I-04	0,7353	4,74	0,08913	2,67	0,56	550	15	560	27	598	23	0,63	8	0,0005
072-A-I-09	0,72257	4,60	0,08907	2,48	0,54	550	14	552	25	561	22	0,68	2	0,0009
072-A-I-15	0,75070	4,76	0,09133	2,09	0,44	563	12	569	27	590	25	1,06	4	0,0012
072-A-I-19b	0,71359	4,29	0,08823	2,36	0,55	545	13	547	23	554	20	0,00	2	0,0004
072-A-I-01	0,93941	2,94	0,11007	2,01	0,68	673	14	673	20	671	14	0,95	0	0,0004
072-A-I-16	0,88815	4,07	0,10494	1,85	0,45	643	12	645	26	653	24	0,86	1	0,0003
072-A-I-21	1,18814	3,86	0,13100	1,65	0,43	794	13	795	31	799	28	0,30	1	0,0002
072-A-I-05	1,13452	2,71	0,12177	1,49	0,55	741	11	770	21	855	19	0,62	13	0,0007
072-A-I-07	1,51559	2,95	0,14998	2,26	0,77	901	20	937	28	1022	19	0,37	12	0,0004
072-A-I-11	1,89571	4,13	0,18311	1,92	0,46	1084	21	1080	45	1071	39	0,56	-1	0,0013
072-A-I-12	1,91441	3,36	0,18097	1,66	0,50	1072	18	1086	36	1114	32	0,29	4	0,0000
072-A-I-13	1,25790	4,45	0,13097	2,39	0,54	793	19	827	37	918	34	0,27	14	0,0082
072-A-I-18	1,60946	4,10	0,15956	2,17	0,53	954	21	974	40	1018	35	0,53	6	0,0003
072-A-I-19[a]	1,67807	3,58	0,16466	1,60	0,45	983	16	1000	36	1039	33	0,48	5	0,0004
												0,37	12	0,0004
072-A-I-10	7,66631	2,05	0,40217	1,25	0,61	2179	27	2193	45	2205	36	0,56	-1	0,0013

Table 5. U/Pb zircon data of the alkaline basalt sample (RLP-20) for Rodeio Velho Member obtained by in situ LAM-ICPMS-MC.

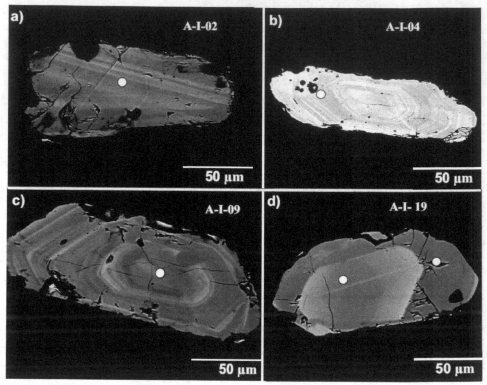

Fig. 11. Zircons microphotographies that defined the 547 ± 6.3 Ma age for Rodeio Velho Member. A-I-09 sample, o = spot showing the place where the analyses were realized in each zircon.

4. Acampamento Velho Formation: geological context, petrography and geochemistry

The earliest references to acid volcanic rocks in Southern Brazil are made by Leinz et al. (1941), who described tuffs, vitrophyres, feldsites and quartz-porphyries, which unconformably overlying the Maricá Formation. This unit was named as Ramada Rhyolite by Robertson (1966) and as Acampamento Velho Formation by Cordani et al. (1974). However, Leites et al. (1990) considered the Acampamento Velho Group as Volcanogenic Sequence III. Paim et al. (2000) named the unit as Acampamento Velho Formation (from Cerro do Bugio Allogroup), and this set of volcanic rocks were considered as an unit limited on the top and base by discontinuities, which marked a special period in the evolution of the basin. It has been traditionally considered as exclusively acid in composition, but detailed geological mapping in the Cerro do Bugio, Cerro do Perau and Serra de Santa Bárbara area (West of Caçapava do Sul town) revealed the existence of a basaltic/andesitic unit at the base and a felsic unit at the top (Zerfass & Almeida 1997, Zerfass et al. 2000, Almeida et al. 2002). This observation leads to the existence of a bimodal alkaline volcanism, mafic at the base and a felsic at the top. Sommer (2000) described the existence of an effusive sequence, pyroclastic and volcanic comenditic rocks at Taquarembó Plateau. In the same area, Wildner et al. (1999) verified that the rocks are alkaline, satured in silica, and have post-collisional characteristics. Sommer et al. (2005) recognized the existence of a bimodal mildly alkaline magmatism related to post-collisional events at Ramada Plateau.

The four main areas of occurrence of the volcanism Acampamento Velho are Santa Bárbara Sub-Basin, Guaritas Sub-Basin and Taquarembó-Ramada Sub-Basin from Camaquã Basin (Paim et al. terminology, 2000 - Fig. 1). At Santa Bárbara and Guaritas sub-basins, the volcanic package is limited at the top and base by sedimentary rocks from own Camaquã Basin, while at Taquarembó-Ramada Basin, the volcanic rocks maintains the relief, sat down on rocks from Sul-Riograndense Shield.

4.1 Geological setting

In the Santa Bárbara area Sub-basin, the area is a long narrow N20°E ridge formed by Acampamento Velho Formation volcanic rocks, where the main elevations are Cerro do Bugio (419 m), Cerro do Perau (331 m) and Serra de Santa Bárbara (440 m), from North to South (Fig. 12). In this area, an unconformity marks the lower contact of the Acampamento Velho Formation over the sedimentary rocks from Maricá or Bom Jardim allogroups (sensu Paim et al. 2000). The upper contact with Santa Fé or Lanceiros alloformations is delineated by a disconformity. The Acampamento Velho Formation is composed of a Lower Mafic Association and an Upper Felsic Association (Fig. 12). The Lower Mafic Association is composed of basaltic and andesitic basalt flows, as well as subordinate andesitic breccias that occur as a continuous bed, with thickness between 10 m and 350 m. It is usually massive, with rare stratification, dipping about 20° to the E or SE (Fig. 12). These rocks show porphyritic texture with plagioclase phenocrysts (Almeida et al. 2002, 2003a).

The Upper Felsic Association is composed of rhyolitic rocks and comprises alternating pyroclastic rocks (lapilli-tuffs, tuffs, welded tuffs) at the base and flows at the top (Fig. 12). Its stratification is tilted, dipping about 20° to the E or SE. The lapilli tuffs are preserved as discontinuous strata of thicknesses up to 40 m. The tuffs occur as lenses of variable thickness (up to 30 m) and internally consist of parallel layers, poorly sorted in general terms. The welded tuffs are also poorly sorted and present predominantly ash fraction, occurring as lenticular layers up to 350 m thick. The rhyolitic flows form a continuous layer of variable thickness from 20 m to 600 m. They display internally of flow foliation, which is frequently folded (Zerfass & Almeida, 1997, 2001). The lapilli-tuffs, tuffs and welded tuffs are interfingered and associated of pyroclastic flows generated during the rhyolitic eruptive phase, as a product of the eruptive column collapse. Pyroclastic fall processes are predominant in distal regions, as it is suggested by the well sorting of the finer tuff members. The rhyolitic flows overlie all of the previous facies, suggesting that the UFA is related to plinian volcanism (Zerfass et al., 2000).

4.2 Mineralogy and petrography

The basalts and andesitic basalts from lower mafic association show relict pilotaxitic texture and zoned plagioclase phenocrysts with diffuse appearance. Sericite, kaolinite, carbonate, chlorite and opaque minerals replace totally the pyroxene phenocrysts and sometimes, partially the plagioclase. Quartz, plagioclase and sanidine grains present on the top layers incipient "kidney-shaped" texture. The matrix is formed by plagioclase microliths, chlorite-carbonate, a ferro-magnesian pseudomorph (pyroxene?) and a large quantity of opaque minerals (Almeida et al., 2002, 2003a). Electron microprobe analyses show that plagioclase phenocrysts and microlites are sometimes totally albitized, with compositions between

$Ab_{99.6}$ and Ab_{98} (Table 6), which should be related to the interaction of late fluids on the plagioclase resulting in the albite formation. The smaller grains in the matrix are strongly altered to kaolinite. The pyroxene is altered to chlorite and generated opaque minerals. The late magmatic fluids rich in CO_2 used part of the Ca from plagioclase and/or pyroxene to form calcite. Albite and clay minerals are therefore product of late fluids relatively enriched in Na. Zircons show prismatic-pyramidal shapes and diffuse zonation (Fig. 9). They occur as inclusions in Ti-rich magnetite and pyroxene (Almeida et al., 2007).

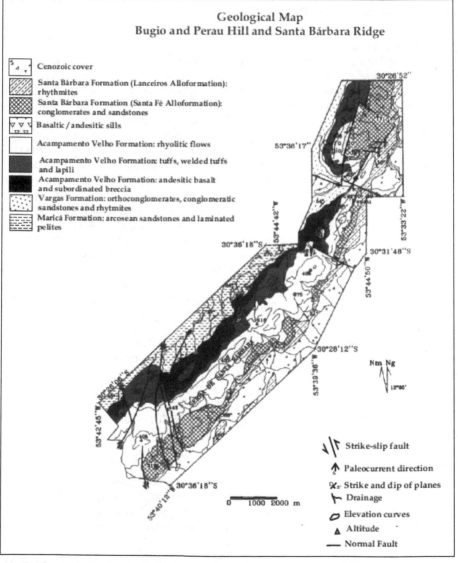

Fig. 12. Geological map of the Cerro Bugio, Cerro Perau and Sierra de Santa Bárbara areas (Almeida et al. 2002).

According to Almeida et al. (2007), for the lower mafic association, the crystallization sequence is: (1) zircon, (2) Ti-rich magnetite, (3) pyroxene, (4) plagioclase, (5) albite resulting from the late introduction of Na- and CO_2-rich fluids, which affected mainly the plagioclase and pyroxene (subordinate), (6) sericite, chlorite and calcite from late solutions that altered the pyroxene, and sericite-kaolinite that altered the plagioclase.

In the Upper Felsic Association, the lapilli-tuffs contain poorly sorted lithoclasts (3 to 40 mm in diameter); vitroclast pseudomorphs (cuspate and platy shapes) substituted by silica and phyllosilicates, quartz crystalloclasts with corrosion gulfs, sanidine and heterogeneous alkali-feldspar (Fig. 13a). The latter is produced by the sodic metassomatic alteration of sanidine, forming heterogeneous pseudomorphs, where part of sanidine is transformed to albite. The lapilli-tuffs matrix is tuffaceous and microcrystalline. The tuffs and welded tuffs differ from each other on the welding degree. They contain crystalloclasts of euhedral quartz or with corrosion gulfs, heterogeneous alkali-feldspar, sanidine (altered to phyllosilicates) and magnetite. Eutaxitic flow structures, conchoidal fractures and perlitic textures are common (Fig. 13b). The tuffaceous matrix is composed of cuspate and platy-shaped fragments (pseudomorphs of volcanic glass shards) and pumice shard-shaped fragments in the welded tuffs (fig. 13c), suggesting pumice pseudomorphs. Original glass is strongly devitrified. Spherical spherulitic structures occur subordinately.

	phenocrysts			matrix		matrix		phenocrysts			matrix
Sample	MH13-10	MH13-12	MH13-13	MH13-14	MH13-15	MH13-16a	MH13-16b	MH13-1a	MH13-6a	MH13-6b	MH13-8
Location	rim	core	core	core	rim	core	rim	core	core	rim	core
SiO_2	71.05	71.65	70.52	71.24	71.23	70.42	71.60	71.37	71.20	70.77	71.51
Al_2O_3	19.97	19.65	20.20	19.84	19.96	20.08	20.18	19.88	20.22	20.38	19.74
FeO	0.35	0.09	0.12	0.05	0.03	0.14	0.22	0.04	0.58	0.21	0.31
MgO	0.03	n.d.	n.d.	n.d.	n.d.	0.01	0.04	0.01	0.10	0.01	n.d.
BaO	0.06	0.08	0.07	n.d.	n.d.	n.d.	0.01	0.07	n.d.	0.03	0.17
CaO	0.13	0.04	0.08	0.09	0.07	0.03	0.06	0.04	0.05	0.25	0.07
Na_2O	10.25	10.34	10.43	10.45	10.09	9.99	10.35	10.64	10.40	9.89	10.14
K_2O	0.10	0.02	0.17	0.03	0.05	0.09	0.02	0.04	0.09	0.11	0.13
Total	101.94	101.87	101.59	101.70	101.43	100.76	102.48	102.09	102.64	101.65	102.07
Si	6.05	6.50	6.03	6.47	6.07	6.05	6.05	6.47	6.03	6.03	6.50
Al	2.00	2.10	2.03	2.12	2.00	2.03	2.01	2.12	2.02	2.05	2.11
Fe_3	0.02	0.01	0.01	n.d.	n.d.	0.01	0.02	0.01	0.04	0.02	0.02
Mg	n.d.	n.d.	n.d.	n.d.	n.d.	n.d.	0.01	n.d.	0.01	n.d.	n.d.
Ba	n.d.	n.d.	n.d.	n.d.	n.d.	n.d.	n.d.	n.d.	n.d.	n.d.	0.01
Ca	0.01	n.d.	0.01	0.01	0.01	n.d.	0.01	n.d.	0.01	0.02	0.01
Na	1.69	1.82	1.73	1.84	1.67	1.66	1.70	1.87	1.71	1.64	1.78
K	0.01	n.d.	0.02	n.d.	0.01	0.01	n.d.	n.d.	0.01	0.01	0.02
Cations	9.80	10.43	9.83	10.45	9.76	9.77	9.79	10.47	9.82	9.77	10.43
X	8.05	8.59	8.06	8.59	8.08	8.08	8.06	8.59	8.05	8.08	8.59
Z	1.75	1.83	1.77	1.86	1.68	1.69	1.72	1.89	1.77	1.69	1.83
Ab	98.70	99.60	98.50	99.30	99.20	99.20	99.50	99.60	99.20	98.00	98.70
An	0.70	0.20	0.40	0.50	0.40	0.20	0.40	0.20	0.20	1.30	0.40
Or	0.64	0.16	1.08	0.22	0.36	0.60	0.12	0.21	0.58	0.72	0.89

Table 6. Albitized plagioclase analysis (electron microprobe) of the Lower Mafic Association from basaltic and andesitic basalt flows from Acampamento Velho Formation. n.d. = not detected. (Almeida et al., 2007).

Tuff analyses by electron microprobe (Table 7) show that the alkali-feldspar crystalloclasts are totally albitized ($Ab_{99.6}$ to $Ab_{98.9}$). Scanning electron microscope analysis show that alkali-feldspar is heterogeneous, contain albite and sanidine in the same crystals. Albite is the product of sanidine alteration. These tuffs display shard pseudomorphs devitrified to illite (Fig. 14a) and crystalloclasts of sanidine with illite pseudomorphs (Fig. 14b). Magnetite is Ti-rich (Table 8) and has inclusions of zircon grains, which are also located around the grain edges. Ti- rich magnetite is pseudomorphically replaced by sanidine and ilmenite along cleavages planes, and their edges are partially corroded by reaction with the matrix (Fig. 14c).

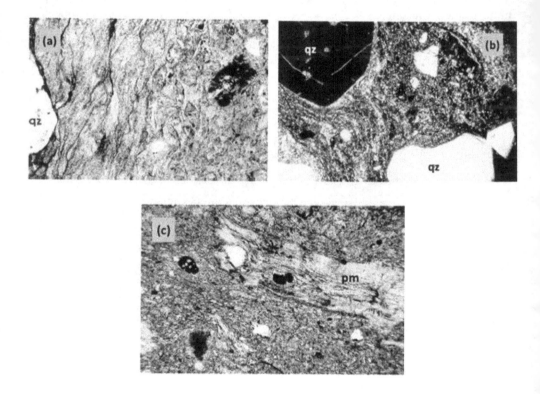

Fig. 13. Tuff photomicrography - 40x- optical microscope PL (a) – the tuff shows cuspate and platy-shaped fragments, volcanic shards pseudomorphs, quartz (qz) and sanidine (sa) crystalloclasts; (b) – welded tuff with eutaxitic texture. qz = quartz crystalloclast; (c) – welded tuff with pumice lithoclasts (pm), fiammes and eutaxitic texture.

Sample	MH16-1a	MH16-1b	MH16-3a	MH16-3b	MH16-5[a]	MH16-5b
Location	core	rim	core	rim	core	Rim
SiO_2	71.10	71.80	70.85	71.63	71.06	71.06
Al_2O_2	19.96	20.13	20.10	20.07	20.12	20.12
FeO	0.05	n.d	0.04	n.d	n.d	n.d
BaO	0.10	0.01	n.d	0.10	0.04	0.04
CaO	0.14	0.05	0.02	0.04	0.04	0.04
Na_2O	10.07	9.96	10.20	10.40	10.63	10.63
K_2O	0.06	0.11	0.03	0.06	0.04	0.04
Total	101.48	102.06	101.24	102.31	101.94	101.94
Si	6.066	6.08	6.05	6.06	6.04	6.04
Al	2.01	2.01	2.02	2.00	2.01	2.01
Fe_2	n.d	n.d	n.d	n.d	n.d	n.d
Ba	n.d	n.d	n.d	n.d	n.d	n.d
Ca	0.01	n.d	n.d	n.d	n.d	n.d
Na	1.17	1.64	1.69	1.71	1.75	1.75
K	0.01	0.01	n.d	0.01	0.01	0.01
Cations	9.77	9.74	9.78	9.79	9.82	9.82
X	8.07	8.09	8.08	8.07	8.06	8.06
Z	1.69	1.65	1.70	1.72	1.76	1.76
Ab	98.90	99.00	99.60	99.40	99.50	99.50
An	0.70	0.20	0.10	0.20	0.20	0.20
Or	0.36	0.73	0.24	0.41	0.28	0.28

Table 7. Albite analyses (electron microprobe) of the Upper felsic Association –
Acampamento Velho Alloformation rhyolitic tuffs. n.d. = not detected. (Almeida et al.,
2007).

The analyses of welded tuffs by electron microprobe (Table 9) show that the alkali-
feldspar crystalloclasts are composed predominantly by sanidine with variable amounts
of albite and K-sanidine (Fig. 15a). The matrix contains predominantly K-sanidine and
also sanidine Na-rich (Ab=32). Plagioclase crystalloclasts (andesine-labradorite) are
present in some samples. The welded tuffs present pseudomorph pumices in fiammes,
heterogeneous alkali-feldspar and sanidine crystalloclasts that are altered to illite and

sometimes corroded by matrix (Fig. 15b). The Ti-rich magnetite crystalloclasts are altered to ilmenite and rutile (Table 8), which are disposed according to twinning and/or cleavage planes (Fig. 15c), and sometimes replaced by sanidine. Homogeneous and zoned zircons usually occur as inclusions, similar to those observed in tuffs, in heterogeneous alkali-feldspar and quartz.

Sample	SSB-32-1	SSB-32-1	MH-32-2	MH-35-1	MH-35-2	SSB-23a
rocks	tuff		welded tuff			flow
Mineral	Mt	Mt	Mt	Il	Ru	Mt
SiO_2	9.80	9.21	n.d.	17.58	4.12	0.99
TiO_2	N.d.	n.d.	7.45	35.81	89,83	15.44
Al_2O_3	1.61	1.12	0.47	11.63	2.04	n.d.
Fe_2O_3	87.94	89.10	92.08	33.35	1.73	83.57
MgO	n.d.	n.d.	n.d.	0.26	1.48	n.d.
K_2O	n.d.	n.d.	n.d.	1.36	0.80	n.d.
total1	100.00	99.99	100.00	99.99	100.00	100.00
Total	99.35	99.43	100.00	99.99	100.00	100.00
Cl	0.65	0.56	n.d.	n.d.	n.d.	n.d.
Si	4.85	0.13	n.d.	8.22	1.93	0.46
Al	0.85	0.11	0.25	6.16	1.08	n.d.
Ti	n.d.	n.d.	4.47	35.81	53.58	9.26
Fe_3	61.51	62.32	64.40	23.32	1.21	58.45
Mg	n.d.	n.d.	n.d.	0.16	n.d.	n.d.
K	n.d.	n.d.	n.d.	1.13	0.66	
Cations	67.21	62.56	69.12	74.80	58.73	68.17

Table 8. Oxides analysis (scanning electron microscope) of the Upper Felsic Association – Acampamento Velho Formation tuffs, welded tuffs and flows. Mt = magnetite; Il = ilmenite; Ru = rutile. n.d. = not detected. The Si and Al detected is due probably the sanidine contamination that is inside the Ti-enriched magnetite. (Almeida et al, 2007).

The rhyolitic flows are homogeneous or banded. Relict structures of perlitic devitrification and conchoidal fractures are common in the microfelsitic matrix. Sanidine, heterogeneous alkali-feldspar and quartz phenocrysts display corrosion gulfs and conchoidal fractures (Figs. 16a). Iron oxide/hydroxide and sericite are also present as alteration products. Rhyolitic flows, when banded, show an intercalation of thick spherical spherulites, product of devitrification, and microcrystalline bands of quartz and feldspar. Electron microprobe analyses of spherulites show heterogeneous composition with fine aggregates of anorthoclase and albite grains (Fig. 16b). Analyses of rhyolitic flows by electron microprobe (Table 10) show that the alkali-feldspar phenocrysts consist of sanidine, albite and heterogeneous alkali-feldspar (Figs. 16c and d). Scanning electron microscope analyses indicate that Ti-rich magnetite (Table 8) and zircon are similar to those described before. The Ti-rich magnetite crystalloclasts are pseudomorphically replaced by sanidine, rutile and ilmenite, which upon interaction with a late fluid altered to TiO_2, probably rutile and hematite, according to the reaction: $4FeTiO_3 + O_2 \rightarrow 4TiO_2 + 2Fe_2O_3$. The interaction with this late fluid also promoted the migration of Ti into cleavage planes and the crystallization of ilmenite and rutile.

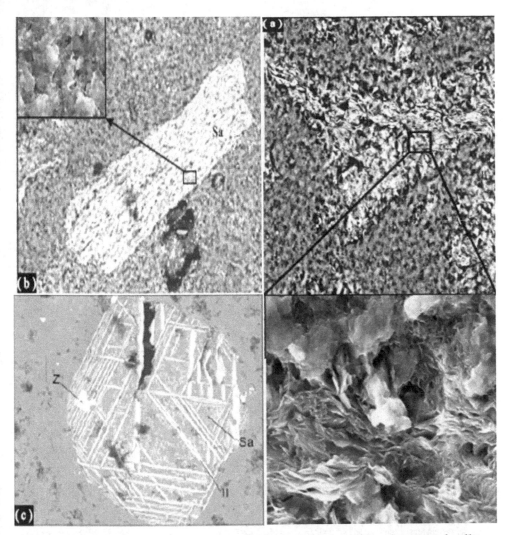

Fig. 14. Tuff: (a) glass shard photomicrography with pseudomorphic substitution by illite (scanning electron microscope); (b) photomicrography of the sanidine altered to illite crystalloclast (scanning electron microscope); (c) Ti-rich magnetite crystalloclast pseudomorphically replaced by ilmenite and sanidine with zircon inclusions (scanning electron microscope).

(a)

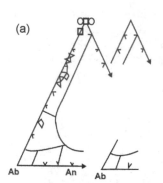

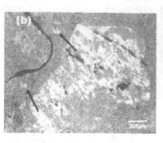

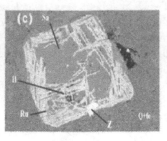

Ab   An   Ab

Fig. 15. Welded tuff: (a) diagram Ab-Or-An (electron microprobe and scanning electron microscope) = matrix, ◆ = rim, Δ = inter, ∇ = core, + = bright, x = dark, O = in Mt. (b) Photomicrography of the heterogeneous alkali feldspar crystalloclast corroded by the matrix (analyzed through the electron microprobe); (c) Photomicrography of the Ti-rich magnetite crystalloclast altered to ilmenite (Il), rutile (Ru) disposed according to twinning and/or cleavages with zircon inclusions (Z) and replaced by sanidine (Sa) (scanning electron microscope).

Sample	MH21-21	MH37-32bb	MH37-35b	MH 37-35c	MH37-38	MH37-44b	MH37-44c	MH37-47b	MH37-49	MH29-p2*	MH29-p2-2*	MH29-p2-2*	MH32-p1*	MH35-1*	MH35-2*
										heterogeneous alkaline feldspars					
Location	matrix	rim	rim	inter	core	rim	inter	rim	core	brigth	dark	brigth	matrix	In Mt	In Mt
SiO2	93.40	84.78	88.03	86.09	95.48	85.10	90.06	97.38	89.14	59.73	68.16	52,01	45.12	67.70	55.12
TiO2	n.d.	n.d.	n.d.	n.d.	n.d.	n.d.	n.d.	n.d.	n.d.	n.d.	n.d.	n.d.	n.d.	n.d.	0.58
Al2O3	3.64	8.92	7.66	7.45	0.18	9.32	6.02	1.16	6.74	14,48	16.30	28.17	23,46	18.57	27.23
Fe2O3	n.d.	n.d.	n.d.	n.d.	n.d.	n.d.	n.d.	n.d.	n.d.	n.d.	n.d.	n.d.	n.d.	3.75	4.31
FeO	0.07	0.12	0.23	0.74	0.15	0.39	0.714	0.05	0.12	n.d.	n.d.	n.d.	19.25	n.d.	n.d.
MgO	0.01	n.d.	0.01	n.d.	0.02	0.01	0.01	n.d.	n.d.	n.d.	n.d.	n.d.	n.d.	0.90	0.950
BaO	n.d.	0.10	0.02	0.03	n.d.	n.d.	0.01	n.d.	0.04	n.d.	n.d.	n.d.	n.d.	n.d.	n.d.
CaO	n.d.	0.06	0.02	0.08	0.01	0.12	0.14	0.01	0.09	n.d.	n.d.	n.d.	n.d.	n.d.	n.d.
Na2O	0.15	3.84	1.49	1.05	0.35	1.78	1.57	0.11	1.17	7.38	6.85	7.29	n.d.	n.d.	n.d.
K2O	2.21	2.41	3.80	4.13	1.49	4.96	3.34	0.34	3.25	18.41	8.69	11.65	12.18	9.07	11.82
Total	99.48	100.29	101.26	99.57	100.68	101.68	101.86	99.05	100.55	100.00	100.00	99.12	100.01	99.99	100.01
Si	15.82	14.66	15.00	14.94	15.91	14.64	15.23	16.30	15.19	27.92	31.86	24.31	21.09	31.65	25.76
Al	0.73	1.82	1.54	1.53	0.62	1.87	1.19	0.23	1.35	7.66	8.63	14.91	12.42	9.83	14.41
Fe3	n.d.	n.d.	n.d.	n.d.	n.d.	n.d.	n.d.	n.d.	n.d.	n.d.	n.d.	n.d.	13.46	2.62	3.01
Ti	n.d.	n.d.	n.d.	n.d.	n.d.	n.d.	n.d.	n.d.	n.d.	n.d.	n.d.	n.d.	n.d.	n.d.	0.35
Fe2	0.01	0.12	0.03	0.11	0.02	0.05	0.10	0.01	0.02	n.d.	n.d.	n.d.	n.d.	n.d.	n.d.
Mg	n.d.	n.d.	n.d.	n.d.	n.d.	n.d.	n.d.	0.01	n.d.	n.d.	n.d.	0.53	n.d.	0.55	0.57
Ba	n.d.	0.01	n.d.	0.01	n.d.	n.d.	n.d.	n.d.	n.d.	n.d.	n.d.	n.d.	n.d.	n.d.	n.d.
Ca	n.d.	0.01	n.d.	0.02	n.d.	0.02	0.03	0.01	0.02	n.d.	n.d.	n.d.	n.d.	n.d.	n.d.
Na	0.05	1.29	0.49	0.35	0.11	0.59	0.51	0.04	0.39	n.d.	5.58	n.d.	n.d.	n.d.	n.d.
K	0.48	0.53	0.82	0.92	0.32	1.08	0.72	0.07	0.71	15.28	n.d.	9.67	10.11	7.53	9.81
Cations	17.08	18.34	17.82	17.90	16.99	18.26	17.78	16.64	17.68	50.86	45.57	49.42	57.08	52.18	53.91
X	16.54	16.48	16.53	16.50	16.53	16.51	16.42	16.52	16.55	35.58	40.49	39.22	46.97	44.10	43.53
Z	0.54	1.85	1.36	1.39	0.46	1.75	1.36	n.d.	1.13	15.28	5.08	10.20	10.11	8.08	10.38
Ab	9.30	70.30	37.20	27.50	26,20	34.90	40.80	32.40	35.00	n.d.	100.00	n.d.	n.d.	n.d.	n.d.
An	n.d.	0.60	0.30	1.20	0.70	1.30	2.10	2.70	1.40	n.d.	n.d.	n.d.	n.d.	n.d.	n.d.
Or	90.70	29.06	62.50	71.28	73.15	63,82	57.09	64.86	63.60	100.00	n.d.	100.00	100.00	100.00	100.00

Table 9. Sanidine and heterogeneous alkaline feldspar analysis from Upper felsic association – Acampamento Velho Formation welded tuffs. Data obtained through the electron microprobe and scanning electron microscope: inter = intermediary; In Mt = analyses made in magnetite that is substituted by sanidine of the matrix. n.d. = not detected. (Almeida et al., 2007).

Almeida et al. (2007) mentioned for felsic rocks the crystallization sequence: (1) zircon, (2) Ti-rich magnetite, (3) sanidine and (4) quartz. The introduction of late Na-rich fluids generated the formation of (5) heterogeneous alkali-feldspar, (6) ilmenite and rutile from the Ti-rich magnetite, (7) albite in the spherulites and finally generated, the alteration of sanidine, vitroclasts and pumice to (8) illite.

4.3 Geochemistry of major, trace and Rare-Earth elements

Almeida et al. (2002, 2003a) already published the Acampamento Velho Formation geochemical characteristics and the text above is a synthesis of mentioned papers.

Twenty samples were analyzed for major, trace and Rare-Earth elements using Argonium Plasma Spectrometry (ICP) at Activation Laboratories LTD (ACTLABS, Canada). The results shown in Table 11 correspond to four samples of the basalts and andesitic basalts, two of the tuffs, four of the welded tuffs and ten of the rhyolitic flows.

For the Lower Mafic Association group (Table 11), whole-rock analyses show a SiO_2 average content of 49.5 wt%; Na_2O = 4.3 wt%, K_2O = 0.8 wt% and CaO = 3.1 wt%. The REE behaviour points out to a moderate alkaline character with high La/Yb_N ratios (5.3 < La/Yb_N < 7.4, average of 6.2) and Eu/Sm_N ratios (0.7 < Eu/Sm_N < 0.8, average of 0.8). The LREE patterns show relatively low fractionation (2.2 < La/Sm_N < 3.1, average of 2.7) with a very slight negative Eu anomaly (0.9 < Eu_N/Eu^* < 0.8, average of 0. 9) (Table 11, Fig. 17a).

Sample	123a2a	123a2b	123a2c	123a2d	333b1a	333b1b	33b2a	333b2b	69.2-1a	69.2-1b	69.2-2a	69.2-2b	ps60-1a	ps60-3
Location	bright core	dark core	bright rim	dark rim	core	rim	core	core	core	rim	core	rim	core	matrix
	heterogeneous alkaline feldpars				sanidine		albita		albita		albita		sanidine	albita
SiO_2	64.12	67.78	64.57	67.78	64.08	64.08	86.11	84.45	68.65	68.57	67.99	68.29	64.51	67.56
Al_2O_3	18.31	19.72	18.47	19.10	18.26	18.43	8.36	9.56	19.54	19.77	19.52	19.35	18.40	19.45
FeO	n.d.	n.d.	n.d.	n.d.	n.d.	n.d.	n.d.	0.67	n.d.	n.d.	n.d.	n.d.	n.d.	0.67
Na_2O	0.35	11.71	0.28	11.21	0.45	0.25	4.33	4.68	11.60	12.10	11.77	11.64	n.d.	11.24
K_2O	16.36	0.18	16.6	1.11	16.32	16.73	0.43	1.39	n.d.	n.d.	n.d.	0.08	19.72	1.48
Total	99.14	99.39	99.92	99.20	99.11	99.49	99.23	100.43	99.79	100.35	99.28	99.36	99.63	100.37
Si	2.99	2.98	2.99	3.00	2.99	2.99	4.50	4.41	3.00	2.99	2.99	3.00	3.00	2.97
Al	1.01	1.02	1.01	1.00	1.01	1.01	0.52	0.59	1.01	1.02	1.01	1.00	1.01	1.01
Fe_2	n.d.	n.d.	n.d.	n.d.	n.d.	n.d.	n.d.	0.02	n.d.	n.d.	n.d.	n.d.	n.d.	0.02
Na	0.03	1.00	0.03	0.98	0.04	0.02	0.44	0.47	0.98	1.01	1.00	0.99	n.d.	0.96
K	0.97	0.01	0.98	0.06	0.97	1.00	0.03	0.09	n.d.	n.d.	n.d.	n.d.	0.99	0.08
Cations	5.00	50.10	5.01	5.04	5.01	5.02	5.49	5.58	4.99	5.02	5.00	4.99	5.00	5.04
X	4.00	4.00	4.00	4.00	4.00	4.00	5.02	5.00	4.01	4.01	4.00	4.00	4.01	3.98
Z	1.00	1.01	1.01	1.04	1.01	1.02	0.47	0.58	0.98	1.01	1.00	0.99	0.99	1.06
Ab	3.00	99.00	3.00	94.20	4.00	2.00	93.60	83.90	100.00	100.00	100	100	n.d.	92.30
An	n.d.	n.d.	n.d.	n.d.	n.d.	n.d.	n.d.	n.d.	n.d.	n.d.	n.d.	n.d.	n.d.	n.d.
Or	97.00	0.99	97.03	5.77	96.04	98.04	6.38	16.07	n.d.	n.d.	n.d.	n.d.	100.00	7.69

Table 10. Sanidine (electron microprobe), albite and heterogeneous alkaline feldspar analysis from Upper Felsic Association – Acampamento Velho Formation rhyolitic lavas. n.d. = not detected. (Almeida et al., 2007).

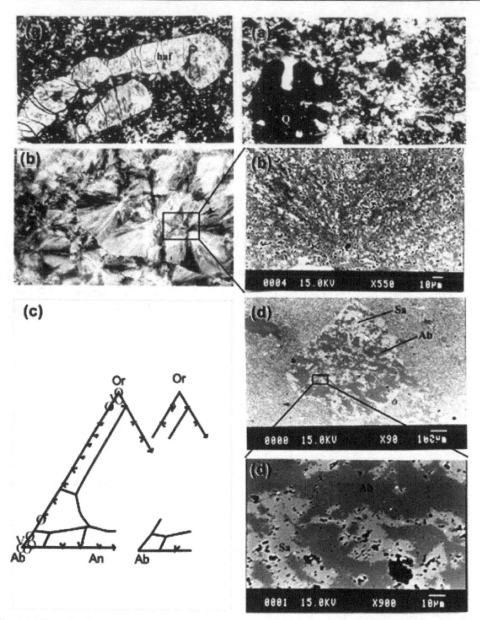

Fig. 16. Rhyolite flows: (a) Photomicrography of heterogeneous alkali-feldspar (haf) and quartz (Q) phenocrysts with gulfs of corrosion and conchoidal fractures (optical microscope PL, 40 x); (b) Photomicrography of spherulites (optical microscope) with fine aggregates of anorthoclase and albite grains as well as Fe oxides (electron microprobe); (c) Ab-Or-An diagram (electron microprobe), O = core, x = rim, + = matrix, ◊ = bright core, ∇ = dark core, □ = bright rim, ∆ = dark rim; (d) Photomicrography of heterogeneous alkali-feldspar phenocryst and its detailed view (electron microprobe).

For the Upper Felsic Association, the tuff samples show a SiO_2 average of 73 wt%, with a low alkalinity (average of Na_2O = 1.5 wt% and K_2O = 3.9 wt%), and high CaO content (2.9 wt%). The light Rare Earth element pattern shows slight fractionation (1.6 < La/Sm_N < 7.9, average of 4.7), with a variable Eu negative anomaly (Eu_N/Eu^* = 0.1 to 0.3, average of 0.2). The welded tuffs are also highly siliceous with SiO_2 average of 78 wt%, low CaO content (average of 0.1 wt%) and the alkalinity higher than tuffs (Na_2O average = 1.7 wt% and K_2O = 5.7 wt%). The Rare Earth element behaviour is similar to the tuffs, although they present much more pronounced light Rare Earth element fractionation (4.1< La/Sm_N < 21.2, average of 7.2) and an important Eu negative anomaly (0.1 < Eu_N/Eu^* < 0.2, average of 0.1) (Table 11, Fig. 17b).

Likewise, the rhyolitic flow samples are also siliceous, with SiO_2 average of 77 wt%, low CaO content (average of 0.2 wt%) and normal alkalinity, although these rocks are more sodic than the pyroclastic ones (Na_2O average = 2.2 wt% and K_2O = 5.3 wt%). The REE pattern is similar to the tuffs and welded tuffs, with LREE fractionation (1.7 < La/Sm_N < 12.2, average of 4.5) and Eu negative anomaly (0.1 < Eu_N/Eu^* < 0.3, average 0.1) (Table 11, Fig. 17b).

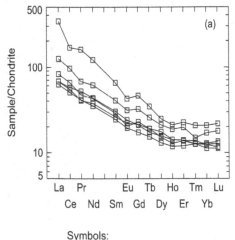 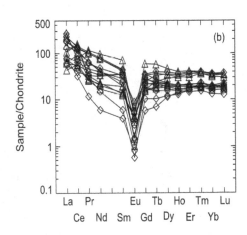

Symbols:

△ Flows
◇ Tuffs ⟩ Upper Felsic Association

□ Flows: Lower Mafic Association

Fig. 17. Chondrite diagram normalized by Rare Earth element (Taylor and MacLennan, 1985) (Almeida *et al.* 2002) for the Acampamento Velho Formation showing a similar pattern for both associations, except for the strong negative Eu anomaly in the Upper Felsic Association rocks. (a) Lower Mafic Association; (b) Upper Felsic Association. (Almeida et al., 2002, 2003a).

	Basalt – Andesitic basalt					Tuff		Welded tuff			
	MH9	MH13	MH14	SSB-7	SSB-32B	MH15	SSB-14	MH31	MH37	MH41	SSB-13
SiO_2	47.05	51.48	54.31	48.84	45.72	78.54	67.64	77.46	77.05	76.59	81.73
TiO_2	1.57	1.53	2.19	1.41	1.88	0.09	0.44	0.12	0.11	0.14	0.11
Al_2O_3	16.49	15.59	12.12	15.19	14.75	9.39	11.45	11.3	11.47	11.49	9.08
Fe_2O_3	8.84	9.13	6.44	9.73	10.63	0.95	3.29	1.87	1.56	2.22	2.08
MnO	0.12	0.07	0.17	0.15	0.15	0.03	0.07	0.01	0.01	0.01	0.01
MgO	5.57	4.23	2.14	4.2	6.19	0.28	0.95	0.08	0.09	0.13	0.03
vCaO	6.31	1.29	5.71	4.96	8.05	1.94	4.03	0.08	0.09	0.05	0.03
Na_2O	4.38	5	4.5	4.93	2.71	1.97	1.1	2.26	1.78	2.41	0.4
K_2O	0.77	0.63	1.2	1.34	0.35	3.74	4.18	5.56	5.5	5.54	6.42
P_2O_5	0.27	0.28	0.41	0.26	0.41	0.01	0.1	0.02	0.02	0.04	0.04
LOI	8.84	7.62	9.29	8.84	9.58	2.81	6.26	1.07	1.34	1.31	1.04
Total	99.81	99.85	99.49	99.84	100.43	99.77	99.51	99.91	99.02	100.1	100.96
Ba	240	174.9	353.3	203	355	225	270	92.8	158.9	137.8	247
Rb	12.1	13.8	26	30	29	158.3	167	101.7	97.2	91.6	141
Sr	519	279.7	225.7	111	555	86.6	87	33.8	39.7	45.4	41
Ta	0.58	0.8	0.86	0.49	0.55	1.9	1.13	1.09	1.08	1.05	1.29
Nb	9.6	11.5	16.2	6	10	17.7	15	17.6	18	17.1	6
Hf	5	5.5	7.9	4.3	4.8	4.2	6.5	7.2	7.2	7.8	7.5
Zr	207.1	231.1	330.3	169	197	127.2	191	207.5	233	236.5	134
Y	31	32	42	28	32	91	38	30	35	36	13
Th	3.45	3.03	4.19	2.7	1.5	11.21	9.7	9.37	6.32	9.1	11
La	25.1	30.6	45.9	23	25	21.3	97	47.3	28.5	64	66
Ce	52.2	63	92.3	48	55	37.8	72	127.2	30.9	113.4	121
Pr	5.54	6.45	9.33	5.64	7.09	4.52	8.24	4.28	1.6	8.86	13.5
Nd	26.5	30.6	43.6	25	31	22.6	34	10.5	4.2	29.1	54
Sm	6	6.6	9.4	5.6	7	8.1	7.7	1.4	0.9	5.5	10
Eu	1.93	1.77	2.73	1.7	2.09	0.25	0.77	0.05	0.07	0.13	0.47
Gd	6.5	7	9.9	5.3	6.4	10.7	6.9	3.2	1.8	5.2	8.1
Tb	1.1	1.1	1.5	0.9	1	2.3	1.2	0.6	0.4	0.9	1.4
Dy	6.7	5.8	8	5	6	14.7	6.4	4.6	4.3	5.9	8.2
Ho	1.2	1.1	1.6	1	1.2	3.1	1.3	1.1	1.2	1.3	1.7
Er	3.5	3.4	4.9	3	3.5	9.6	3.6	3.9	4	4.4	5
Tm	0.45	0.45	0.53	0.45	0.49	1.42	0.55	0.63	0.82	0.67	0.82
Yb	3.2	3.1	4.2	2.8	3.1	9.1	3.3	4	4.4	4.4	4.9
Lu	0.51	0.49	0.68	0.43	0.44	1.41	0.49	0.84	0.69	0.71	0.77

Table 11. Acampamento Velho Formation. Geochemical analyses of the major, trace and Rare Earth elements (Almeida *et al.*, 2002). These analyses were performed at Activation Laboratories (ACTLAB), Canada, using the Argonium Plasma Spectrometry - ICP. MH = Cerro do Bugio and Perau samples; SSB = Serra de Santa Bárbara samples. Major elements values are in percent (wt%); trace and Rare Earth elements are in part per million (ppm).

	Rhyolitic Flows									
	MH20	MH21B	MH22	MH25	MH27	MH50	MH53	MH60	MH7	SSB-1B
SiO_2	78.97	80.6	78.15	79.26	76.26	75.76	76.85	76.37	77.26	77.79
TiO_2	0.07	0.07	0.09	0.09	0.09	0.12	0.13	0.13	0.12	0.14
Al_2O_3	10.9	8.52	10.07	11.48	11.32	11.73	12.29	12.34	11.67	11.77
Fe_2O_3	1.84	1.49	2.19	1.56	1.67	1.59	1.69	1.48	1.7	2.77
MnO	0.01	0.01	0	0.01	0.01	0.02	0.02	0.02	0.02	-0.01
MgO	0.08	0.02	0.01	0.11	0.06	0.11	0.08	0.1	0.1	0.06
vCaO	0.03	0.02	0.01	0.21	0.12	0.96	0.06	0.17	0.06	0.06
Na_2O	2.98	0.26	0.34	3.32	3.76	2.53	2.7	2.5	0.14	0.81
K_2O	4.18	6.41	7.87	3.9	4.75	5.26	5.49	5.36	5.01	5.37
P_2O_5	0.01	0.02	0.01	0	0	0.01	0.02	0.01	0.03	0.02
LOI	0.68	0.94	0.76	0.75	0.41	1.45	1.12	1.33	2.8	1.9
Total	99.76	98.37	99.51	100.7	98.45	99.56	100.1	99	98.92	100.7
Ba	60.3	72.3	52	139.4	128.6	312.8	428.1	369.1	136.2	190
Rb	100.3	140.5	176.7	127.6	148.3	149	128.4	144.6	119.7	141
Sr	35.8	49.5	36.5	40.3	28	34.9	54.1	52.7	23.2	43
Ta	2.03	1.68	1.63	2.23	2.23	1.47	1.48	1.38	1.38	1.99
Nb	20	26	28.4	13	18	21	21.1	12	20	24
Hf	11.5	7.9	9.9	13.3	13.9	7.8	8.1	7.3	6.9	16
Zr	172	270	343.4	196	233	276	253.1	189.2	254	601
Y	45	55	85	114	35	57	46	42	46	75
Th	14.33	15.55	12	16.6	14.53	13.98	13.07	13.28	12.73	15
La	26.4	30.8	54.4	98.1	16	81.4	39	23.8	85.2	22
Ce	48.3	62.3	99.7	72.9	54.9	136	151.6	85.9	127.6	53
Pr	3.55	6.27	9.25	15.98	3.64	14.34	4.55	3.87	13.65	5.64
Nd	12	28.3	30.3	68	15.1	57.4	16.5	14.5	53.4	24
Sm	2.9	6.5	4.6	16.6	4.4	10.5	3.1	3.2	9.5	8
Eu	0.09	0.13	0.14	0.38	0.13	0.62	0.35	0.36	0.5	0.29
Gd	4.7	6.5	7.4	17.9	5	10.8	5.4	4.7	8.8	10
Tb	1.1	1.3	1.9	3.3	1.4	1.6	1	1	1.3	2.1
Dy	7.7	8.1	13.4	17.6	10.9	8.5	7.1	7	7.2	13
Ho	1.6	1.7	2.9	3.4	2.6	1.7	1.6	1.6	1.4	2.6
Er	4.8	5.2	9.1	10.1	8.8	5.4	4.8	4.8	4.5	7.5
Tm	0.65	0.79	1.3	1.42	1.36	0.76	0.7	0.68	0.64	1.28
Yb	4.4	4.9	8.2	8.9	8.6	5	4.7	4.4	4	7.6
Lu	0.68	0.78	1.28	1.35	1.31	0.84	0.73	0.66	0.62	1.14

Table 11. Continuation.

The light Rare Earth element fractionation in the rhyolitic flows is slighter than that observed at welded tuffs and very similar to the tuffs ($1.7 < La/Sm_N < 12.2$, average of 4.5). The Eu anomaly is similar to that of the welded tuffs (Fig. 17b). The Upper Felsic Association Rare Earth element diagram exhibits values corresponding to the evolved rocks, similar to those of Culler & Graf (1984). The rocks show moderate fractionation and clear parallelism, especially of heavy Rare Earth element and confirm the alkaline character. The increasing values of Rare Earth element and the marked negative Eu anomaly are common in the felsic rocks associated with mafic one. According to Almeida et al. (2002, 2003a), the Acampamento Velho Formation bimodal volcanism is characterized by the presence of dominant acid and subordinated mafic rocks, with overall absence of rocks with SiO_2 content between 54 wt% and 67 wt%. The Nb x Zr and Y x Zr diagrams (Fig. 18) show different evolutionary trends for the mafic and felsic successions, reinforcing the bimodal character of this magmatism.

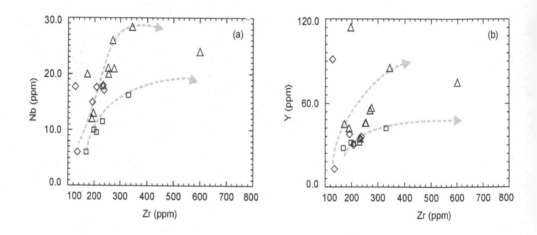

Fig. 18. Pearce and Norry (1979) plotting shows two distinct evolutionary trends for Lower Mafic Association and Upper Felsic Association from Acampamento Velho Formation (Almeida *et al* 2002). (a) Zr versus Nb and (b) Zr versus Y. Symbols are the same as in figure 17.

5. Rodeio Velho Member: Geological context, petrography and geochemistry

The Rodeio Velho Formation (sensu Ribeiro et al., 1966) has been described as bearing at least three vesicular andesite flows, with thickness estimated at 100 m and no evidence of explosive activity. However, it was observed that the Rodeio Velho Member manifested itself as flows, pyroclastic deposits and shallow intrusions. There are few petrographic and geochemical studies on the Rodeio Velho Member. Silva Filho (1996) showed the intrusive character of the magmatism, rejecting former ideas of an exclusively volcanic event. Fragoso Cesar et al. (2000) named these rocks as Rodeio Velho Intrusive Suite, which are represented by tabular intrusions within the sub-horizontal continental deposits of the Guaritas Group. Lopes et al. (1999) mentioned its occurrence in subsurface at the CQP-1-RS probing, North Camaquã Mines, where reach 119.50 thickness. Almeida et al. (2000, 2003) studied this event at the following areas within Camaquã Basin (Fig. 1): (i) Santa Bárbara Sub-basin: in the South-centre and North-centre portions from Arroio Santa Bárbara and Arroio Carajás, (ii)

Guaritas Sub-basin: at the Minas do Camaquã, Rodeio Velho, Passo do Moinho and Pedra da Arara regions and Arroio dos Neves drainage line.

5.1 Geological setting
The studied regions are localized in Southeast, South and Southwest Caçapava do Sul city (RS), in the Camaquã Basin (Fig.1). The text is a synthesis of Almeida et al. (2000, 2003a and 2003b).

5.1.1 Santa Bárbara sub-basin
The Rodeio Velho Member volcanic rocks are localized South-centre and North-centre portions from Arroio Santa Bárbara and Arroio Carajás (topographic map, respectively). This event presents four volcanic cones opens to SE and lined up NNE-SSW direction, parallels to the regional flow directions, evincing the structural control in the positing of these cones. Ejected bombs and blocks of vesicular basaltic andesitic rocks, local concentrations of jasper, chalcedony and geodes are present throughout the area. They probably represent the end of a strombolian-type event associated with the evolution of the cones. The preservation of this structure is regarded as due to the presence of younger sedimentary rocks across the region. These sedimentary rocks include arkosic sandstones, conglomeratic sandstones, and conglomerates. They correspond to the sedimentation named as Varzinha Alloformation (or Formation) from the Guaritas Allogroup (Table 1). An elliptical caldera that shows its longest axis about 7.2 km NNE-SSW and it is found in the surrounding of the cones. (Fig. 19).

To the West and South of the cones, Rodeio Velho Member basaltic dykes intrude the sediments building the Lanceiros Alloformation when they are solidified. These layers of the basaltic rocks, sub-vertical and preferentially aligned to N26ºE, are massive in the centre and vesicular along the borders. They have been interpreted as a Rodeio Velho Member intrusive manifestation. However, considering that a dyke-sandstone contact was found in the sandstones vesicles, which was probably generated by gases coming from the magma and suggests that the sediment was still unconsolidated, so, the Rodeio Velho Member volcanism would be contemporary with the formation of the sandstone from Pedra Pintada Formation. 2 km to the South of the cones, in the Carajás Creek, the Rodeio Velho Member also crops out as lava flows, showing *aa* and *pahoehoe* structures with centimetric to decimetric hollow tubes, as well as amygdales and/or vesicles that are up to 5 mm long at the base and top of each flow (Fig. 20a). The sedimentary rocks layers are mainly composed of middle granulometry to fine sandstones and show crossed stratification of very low angle. Levels of pelites associated to this sandstone show feature of subaerial exposure with contraction cracks (Fig. 20b).

Rodeio Velho Member is positioned as a flow that went through the dunes, when the solidification occurs, the structure looks like an intrusion of the volcanic in the sedimentary rocks, but they correspond to a volcano-sedimentary interaction of the Rodeio Velho Member and Pedra Pintada Formation, and rhythmites from Guaritas Group (Fig. 21a). The volcano-sedimentary interaction features include also flow striation, xenoliths, clastic dykes, marks in increasing and peperites (Figs. 21 b, 21c and 21d). Based on field characteristics, it seems that the emitting centers would be exactly at the cones, which are 2 km North and open towards the South (Almeida et al., 2000).

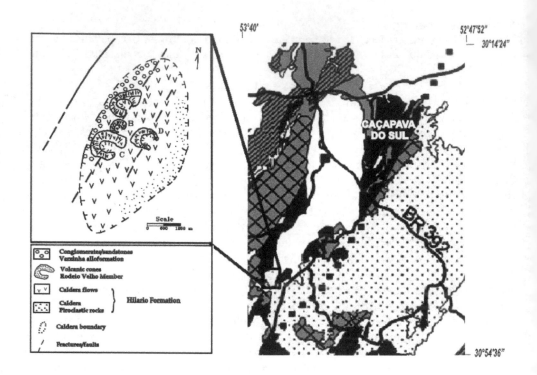

Fig. 19. Map for Camaquã Basin area showing the caldera and cones (Rodeio Velho volcanic rocks) location in the Camaquã Basin context. Modified from Almeida et al., 2003b.

5.1.2 Guaritas sub-basin

The Rodeio Velho Member are located at Minas do Camaquã, Rodeio Velho, Passo do Moinho and Pedra da Arara regions, and Arroio dos Neves drainage line (Fig. 1). These volcanic rocks show flows interdigitated in Pedra Pintada Alloformation rhythmites and the volcano-sedimentary interaction. Field observations indicate a contemporaneity between sedimentation and volcanism. The peperites, flux channels, clastic dykes and other features development suggest interaction between hot lava and wet poorly consolidated sediments. In Pedra de Arara area (Fig. 22a), it is observed the same structure found at Arroio Carajás region with volcanic rock positioned as the flow that went through the dunes, when the solidification occurred. The structure looks like an intrusion of the volcanic in the sedimentary rocks, but they correspond to an volcano-sedimentary interaction in this area with peperites in the sandstone-volcanic rock contact. Immediately above, at Passo do Moinho (South Pedra da Arara, Fig. 1), the sandstones are restricted to the enclaves and clastic dykes (Fig. 22b).

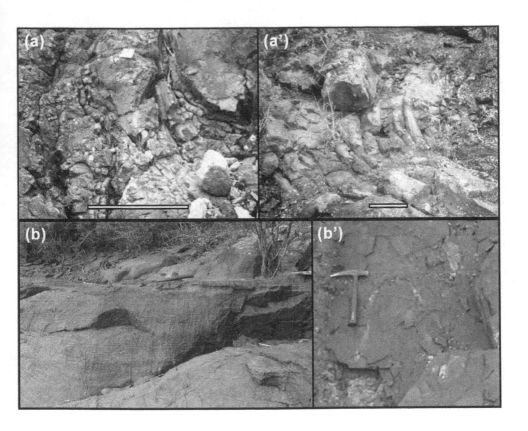

Fig. 20. Carajás Creek: (a) *pahoehoe* lavas downstream of the Arroio Carajás. The scale measures (a) 1 m and (a') 10 cm. (b) - sedimentary lithologies: sandstone with plane-parallel lamination masked by intense fracturing and (b') pelite with contraction cracks (in Petry, 2006).

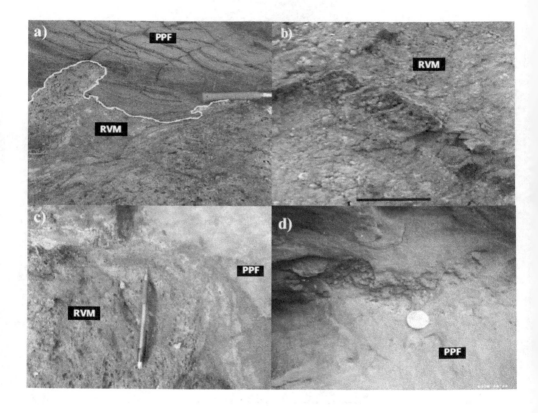

Fig. 21. Volcano-sedimentary interaction features found at Arroio Carajás outcrops – Santa Bárbara Sub-basin: a) volcanic rock (Rodeio Velho Member) and sandstone (Pedra Pintada Formation) positioned as a flow that went through the dunes; b) sandstone clastic dyke in amygdaloidal lava; c) marks in increasing; d) peperite with basaltic clasts evidencing the plasticity.

Fig. 22. Photos showing the several styles of the positioning from Rodeio Velho Member and the volcano-sedimentary interaction features found at Guaritas Sub-basin. a) – volcanic rock (Rodeio Velho Member) and sandstone (Pedra Pintada Formation) positioned as a flow that went through the dunes in Pedra da Arara area. (b) – Detail showing the clastic sandstone dike in amygdaloidal lava at Passo Moinho area.

5.2 Mineralogy and petrography

According to Almeida et al. (2000), the RVM is characterized by mafic and intermediate lava flows, pyroclastic deposits and intrusions. Petrographically, these rocks are andesites, subalkaline basalts and trachyandesites. The rocks show pylotaxitic, vitrophyric or ophitic textures, with glomeroporphyritic euhedric to subedric plagioclase phenocrysts. The plagioclase shows size around 0.5 mm, and is generally altered to carbonate and can be oriented within de matrix. Relics of euhedric pyroxene, olivine and opaques (Ti-magnetite and other Fe-Mn oxides) are present and show sizes that vary from 0.3 to 0.1 mm. Accessory minerals such as apatite and zircon are observed. Intersertal glass is recrystallized (forming spherulites) or altered. The presence of vesicles and amygdales is common and the sizes vary from 0.5 to 10 mm. The amygdales are filled with quartz and carbonate or both of them. The quartz precedes the carbonate. Pyroclastic rocks are stratified and vary from ash-flow tuffs to lapillites, brecciated and poorly sorted. Shards and fiammes attest to the pyroclastic character. Plagioclase and euhedral quartz crystalloclasts (<2%) are dispersed in the tuffaceous, partially glassy matrix.

5.3 Geochemistry

Almeida et al. (2000, 2003a and 2003b) already published the complete Rodeio Velho Member geochemical characteristics. The text above is a synthesis of mentioned papers. The chemical analysis data are shown in Table 12. Considering the high alteration level present in the Rodeio Velho Member rocks, only trace and Rare earth elements were used in the diagrams.

Amostra	PH-3	PH-4	PH5	PM49	PM54	RLP1	RLP10	RLP12	RLP14	RLP15	PLR17
	1	2	3	4	5	6	7	8	9	10	11
SiO_2	50.14	65.53	48.55	51.47	52.93	51.89	57.12	52.33	51.69	54.63	46.84
TiO_2	2.15	1.82	1.71	2.71	2.13	1.66	2.24	1.83	1.856	1.807	2.099
Al_2O_3	14.64	12.7	14.14	15.69	16.2	15.75	14.58	15.77	15.79	15.34	15.77
Fe_2O_3	11.63	6.28	9.77	13.6	13.89	8.92	10.26	8.42	10.61	11.17	12.14
MnO	0.16	0.1	0.17	0.16	0.1	0.26	0.13	0.09	0.07	0.08	0.09
MgO	4.35	0.85	0.34	1.1	1.87	1.66	1.53	4.11	2.96	0	3.68
CaO	7.4	1.99	7.64	2.45	1.04	6.57	3.3	4.97	5.18	2.68	8.26
Na_2O	3.45	6.01	3.34	6.36	7.57	4.59	7.6	3.89	3.78	3.9	3.7
K_2O	1.24	1.06	6.27	1.95	0.88	2.75	0.26	2.24	2.2	5.56	1.43
P_2O_5	1.18	1.2	0.91	1.41	0.36	1.05	1.4	1.08	1.1	1.08	0.85
LOI	3.22	1.42	5.95	2.14	2.35	4.57	1.73	3.89	3.59	2.05	4.51
total	99.56	98.94	98.78	99.05	99.32	99.68	100.15	98.62	98.83	98.83	99.37
Ba	1780	2695	2204	2093	344	1124	673	1150	1142	1560	1140
Rb	15.6	13.9	94.1	28	17	40	4.5	29	28	91	26
Sr	1186	276	222	824	328	558	454	775	808	387	805
Ta	1.36	1.18	1.18	2.1	0.94	1.03	1.67	1.2	1.3	1.2	1.2
Nb	30.2	25.5	27.9	36	19	19	25	27	27	25	20
Hf	7.4	2.5	3.4	8.5	3.3	8	7.8	8.3	7.9	7.5	3.8
Zr	461	399	360	380	185	383	369	395	384	367	164
Y	54	48	36	84	27	51	41	42	41	36	28
Th	3.47	2.61	3.41	3.7	1.4	2.9	3	3.2	3.2	3.1	1.6
La	102.7	86.9	84.5	104	40	136	104	76.6	77.1	80.6	42.8
Ce	204.7	164	168	186	76	217	201	148	145	151	91.3
Pr	20.13	16.53	16.65	23.8	11	26.6	23.8	19.5	19.6	19	11.8
Nd	91.7	75.2	72.7	95	45	108	97	70.4	70	66.8	45.1
Sm	16.2	13.7	12.2	17	8.5	17	16	12	11.7	10.6	8
Eu	4.81	4.02	3.52	4.81	2.52	4.45	4.46	3.58	3.43	3	2.88
Gd	14.5	12.8	11.2	15	6.8	12	11	10.1	9.7	8.7	6.8
Tb	2	1.8	1.4	2.2	1	1.7	1.6	1.5	1.4	1.3	1
Dy	10	8.9	6.8	12	5.5	8.9	7.6	8.1	7.7	6.8	5.8
Ho	1.9	1.8	1.3	2.4	1	1.6	1.4	1.5	1.5	1.3	1.1
Er	5.4	5.2	3.7	7.1	2.9	4.6	3.7	4.3	4	3.6	3
Tm	0.65	0.59	0.45	0.94	0.39	0.61	0.49	0.57	0.56	0.51	0.41
Yb	4.5	3.8	3	5.5	2.4	3.5	2.9	3.5	3.4	3.1	2.5
Lu	0.73	0.6	0.47	0.85	0.34	0.57	0.45	0.52	0.49	0.46	0.36

Table 12. Geochemical analysis of the major, trace and Rare Earth elements (Almeida *et al* 2000). These analysis were performed at Activation Laboratories (ACTLAB), Canada, using the Argonium Plasma Spectrometry - ICP. (1,2, 3) = volcanic cone samples (4, 5) Rodeio Velho flow sample; (6) = Rincão da Tigra flow sample; (7) = flow sample obtained South of Passo Moinho; (8) = Rincão da Tigra flow sample; (9, 10) = Arroio dos Carajás flow sample; (11) = Arroio dos Neves flow sample.

The Rodeio Velho Member rocks were plotted in the Nb/Y versus Zr/TiO_2 classificatory
diagram (Winchester & Floyd, 1977 - Fig. 23a), which allowed classify the flows as sub-
alkaline basalts, andesites, trachyandesites, alkaline basalts and basaltic andesites. The
pyroclastic rocks were classified as trachyandesites, and the epizonal as trachyandesites and
alkaline basalts.

The REE pattern normalized by the chondrite (Nakamura, 1977- Fig. 23b) shows a similar
behavior to the alkaline basalts and the values correspond to the evolved rocks (Culler &
Graf, 1984), displaying a marked fractionation and parallelism between all the rocks.
According to the data and the REE distribution, the light Rare Earth elements (LREE)
enrichment occurs under conditions of low partial melting, especially from a source that
contains garnet, suggesting a great deep for genesis of these rocks.

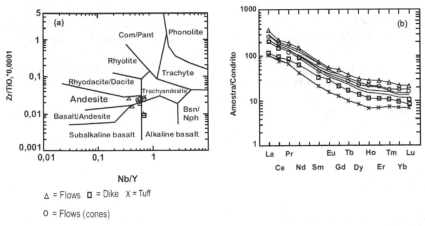

Fig. 23. (a) Zr/TiO_2 *versus* Nb/Y diagram (Winchester & Floyd, 1977) (b) Chondrite diagram
normalized by REE (Nakamura 1977) (Almeida et al., 2000).

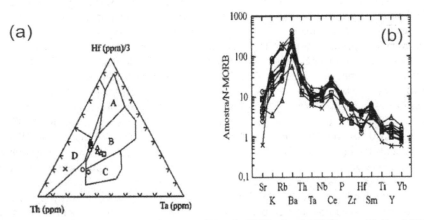

Fig. 24. (a) Tectonic discrimination diagram (Wood, 1980). A = MORB type N; B = MORB
type E, tholeiitic WPB; C = alkaline WPB and differentiated; D = destructive continental
margin basalts and differentiated. Legend for figure 23 (b) Multi-elements spiderdiagram
normalized by Sun and McDonough (1989) values (Almeida et al., 2000).

Considering the chemical data, all the Rodeio Velho Member rocks are similar. Thus, the samples plotted on Hf/3 x Th x Nb/16 diagram (Wood, 1980 - Fig. 24a) show that the rocks were generated in a domain of intraplate continental basalts (alkaline and/or E-MORB type). This behavior is confirmed in the Zr/Y *versus* Zr diagram (Pearce & Norry, 1979) and in the spidergram normalized by N-MORB (Sun & McDonough, 1989 - Fig. 24b). These rocks probably were formed by fractionated crystallization, as indicated by the good correlation between the incompatible elements pairs, such as Ce versus Sm-La, and Zr versus Nb and Y.

6. Rb, Sr, Sm and Nd isotope data for Acampamento Velho Formation and Rodeio Velho Member

Almeida et al. (2005) already published the complete Acampamento Velho Formation and Rodeio Velho Member isotopic characteristics. The text above is a synthesis of the mentioned papers.

The isotopic results obtained in this study are shown in Table 13 and figures 25, 26, 27 and 28. The Acampamento Velho Formation lower mafic association samples show Rb contents between 12 and 83 ppm (average 32 ppm), and Sr contents vary from 23 to 703 ppm (average 309 ppm). Measured $^{87}Sr/^{86}Sr$ ratios in these rocks range from 0.707 to 0.731, while initial ratios, calculated for 550 Ma, are between 0.706 and 0.711. The Sm content varies from 5 to 13.7 ppm, and the Nd ranges between 22.9 and 78.2 ppm. Measured $^{143}Nd/^{144}Nd$ ratios are concentrated between 0.5116 and 0.5119, resulting on $\varepsilon Nd_{(0)}$ from - 9.3 to -16.6, and εNd (t = 550 Ma) of -2.9 to -10.3. TDM model ages lie between 1.11 and 1.78 Ga.

Tuffs and welded tuffs of the upper felsic association display Rb values of 91.6 to 167.0 ppm (average 126 ppm), and 36.2 to 98.8 ppm for Sr (average 55 ppm), compatible with acid volcanics. The measured $^{87}Sr/^{86}Sr$ ratios are high, between 0.745 and 0.771, while the initial values range from 0.701 to 0.713. They present low Sm and Nd contents (0.8 to 9.5 ppm and 4.1 to 43.8 ppm, respectively), and $^{143}Nd/^{144}Nd$ ratios from 0.511 to 0.512. The $\varepsilon Nd_{(0)}$ lies between -11.9 and -16.4, while the εNd at the crystallization time (t = 550 Ma) ranges from -7.2 to -9.8. The TDM model ages for the pyroclastic rocks particulate fraction are 1.3 to 1.9 Ga. The rhyolitic lava flows of the upper felsic association show even higher Rb values (100 to 176 ppm, average 137 ppm) compared to Sr (13 to 55 ppm, average 39 ppm), with very high measured as well as initial $^{87}Sr/^{86}Sr$ ratios, respectively, 0.771 to 0.939 and 0.701 to 0.721. The $^{143}Nd/^{144}Nd$ ratios ranges from 0.511 to 0.512, corresponding to $\varepsilon Nd_{(0)}$ values from -7.14 to -16.25, εNd (for t = 550Ma) from -5.7 to -8.8 and T_{DM} ages between 1.3 and 2.1 Ga.

The Rodeio Velho Member rocks show low Rb (4.5 to 91 ppm, average 31 ppm) and high Sr contents 310 to 1203 ppm, (average 733 ppm), with low Sr isotopic ratios (0.705 to 0.710 measured, and 0.704 to 0.707 for initial $^{87}Sr/^{86}Sr$ calculated for 470 Ma.). The $^{143}Nd/^{144}Nd$ values concentrate between 0.51165 and 0.51193 correspond to strongly negative $\varepsilon Nd(0)$ from -13.89 to -19.36, εNd (t = 470Ma) from -39 to -13.92, and T_{DM} model ages from 1.50 to 1.96 Ga. Plots of initial $^{87}Sr/^{86}Sr$ ratio against $^{143}Nd/^{144}Nd$ and εNd (Fig. 25) show different signatures for Rodeio Velho Member mafic lavas, Acampamento Velho Formation mafic lavas and felsic rocks. The Rodeio Velho Member mafic lavas show a variable radiogenic Nd and $\varepsilon Nd(t)$ to a nearly constant radiogenic Sr (Fig. 25a). The opposite occurs with the Acampamento Velho Formation mafic lavas. The data in figure 25b show a negative correlation between radiogenic Sr and εNd for the Acampamento Velho Formation samples.

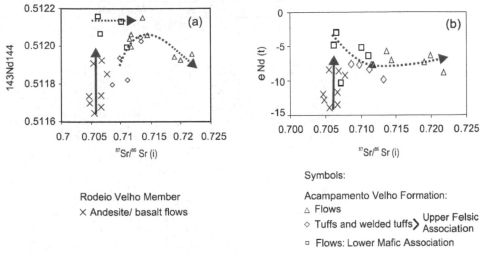

Fig. 25. (a) $^{144}Nd/^{143}Nd$ versus initial $^{87}Sr/^{86}Sr$ and (b) εNd versus initial $^{87}Sr/^{86}Sr$ isotope diagrams for Acampamento Velho Formation and Rodeio Velho Member (Almeida et al., 2005).

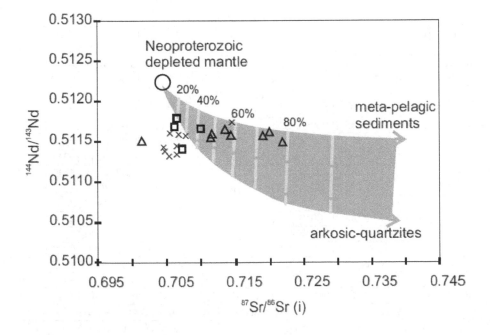

Fig. 26. $^{143}Nd/^{144}Nd$ versus initial $^{87}Sr/^{86}Sr$ diagram for the volcanic rocks from Rodeio Velho Member and Acampamento Velho Formation flows. The plotting shows the field for crustal contamination. Symbols are the same as in figure 25. (Almeida et al., 2005).

Acampamento Velho Formation - Upper Felsic Association								
Sample	SiO_2	Al_2O_3	Ba/Th	Sm (ppm)	Nd (ppm)	$^{143}Nd/$ ^{144}Nd	Erro ppm	$\varepsilon Nd_{(0)}$	$\varepsilon Nd_{(t)}$
#MH-15	78.54	9.39	20.7	7.77	22.65	0.51205	14	-11.56	-12.32
#SSB-14	77.05	11.47	25.1	0.84	4.17	0.51194	15	-11.93	-7.51
##MH-31	77.46	11.3	9.9	1.36	11.44	0.51180	16	-16.44	-7.68
##MH-37	67.64	11.45	27.8	9.47	42.85	0.51203	14	-13.65	-8.36
##MH-41	76.59	11.49	15.1	5.17	28.59	0.51182	12	-16.00	-9.87
##SSB-13	81.73	9.08	22.4	8.35	43.85	0.51198	13	-12.91	-7.18
*MH-17	77.26	11.67	10.7	7.84	35.89	0.51195	12	-13.32	-8.80
*MH-20	78.97	10.9	4.2	2.98	12.40	0.51206	24	-11.24	-7.62
*MH-21b	80.6	8.52	4.6	6.5	28.69	0.51206	11	-11.29	-7.00
*MH-22	78.15	10.07	4.3	4.23	28.87	0.51193	15	-13.90	-6.30
*MH-25	79.26	11.48	8.4	17.27	73.33	0.51215	14	-9.50	-5.69
*MH-27	76.26	11.32	8.8	4.19	14.79	0.51227	18	-7.14	-5.36
*MH-50	75.76	11.73	22.3	10.71	60.40	0.51194	13	-13.57	-7.29
*MH-53	76.85	12.29	32.7	3.07	15.78	0.51200	14	-12.47	-7.72
*MH-60	76.37	12.34	27.7	3.00	13.13	0.51204	14	-11.71	-7.61
*SSB-1b	77.79	11.77	12.6	7.55	24.10	0.51217	13	-9.11	-8.61

(continuation)

Sample	TDM	$^{87}Sr/$ $^{86}Sr_{(m)}$	erro (SD bs)	$^{87}Sr/$ $^{86}Sr_{(i)}$	Rb/Ba	$^{87}Rb/$ ^{86}Sr	erro (SDabs)	$^{147}Nd/$ ^{144}Nd	Erro ppm
#MH-15	---	0.75224	0.00107	0.71338	0.70355	4.95653	0.04957	0.20732	28
#SSB-14	1812.02	0.76577	0.00015	0.71107	0.61170	4.77399	0.04774	0.07200	24
##MH-31	1336.68	0.77196	0.00016	0.70983	1.09590	7.92374	0.07924	0.13358	32
##MH-37	1922.00	0.74597	0.00018	0.70854	0.61851	6.97588	0.06976	0.12139	28
##MH-41	1777.20	0.75648	0.00046	0.71327	0.66473	5.51066	0.05511	0.10928	26
##SSB-13	1638.2	0.7780	0.00038	0.70143	0.57085	9.77349	0.09773	0.11508	24
*MH-17	2021.17	0.77655	---	0.72188	0.87885	6.97300	0.06973	0.13211	48
*MH-20	2171.45	0.77186	0.00065	0.71169	1.66335	7.67411	0.07674	0.14515	22
*MH-21b	1906.70	0.78126	0.00015	0.71438	1.94329	8.52981	0.08530	0.13557	30
*MH-22	1354.44	0.82658	0.00015	0.71997	3.39807	13.59785	0.13598	0.08851	28
*MH-25	1893.04	0.78159	0.00019	0.71367	0.91535	8.6628	0.08663	0.14237	36
*MH-27	2799.16	---	0.00047	---	1.15318	---	---	0.17121	26
*MH-50	1566.44	0.81214	0.00018	0.71899	0.47634	11.88013	0.11880	0.10724	28
*MH-53	1642.65	0.76269	0.00026	0.71176	0.29993	6.49581	0.06496	0.11745	28
*MH-60	2018.02	0.77194	0.00015	0.71149	0.39176	7.71001	0.07710	0.13820	28
*SSB-1b	---	0.93947	0.00015	0.70140	0.74210	30.36370	0.030364	0.18943	26

Acampamento Velho Formation - Lower Mafic Association								
Sample	SiO_2	Al_2O_3	Ba/Th	Sm (ppm)	Nd (ppm)	$^{143}Nd/$ ^{144}Nd	Erro ppm	$\varepsilon Nd_{(0)}$	$\varepsilon Nd_{(t)}$
*MH-15	47.05	16.49	69.5	13.74	78.23	0.511 8	15	-16.66	-10.31
*MH-13	51.48	15.59	57.7	6.26	47.15	0.51207	37	-11.15	-2.97

*MH-14	54.31	12.12	84.3	8.3	40.58	0.51199	14	-12.55	-7.49
*SSB-7	48.84	15.19	75.1	5.02	22.94	0.51216	13	-9.34	-4.82
*SSB-32b	45.72	14.75	236.6	6.81	31.69	0.51213	13	-9.86	-5.17

(continuation)

Sample	TDM	$^{87}Sr/$ $^{86}Sr_{(m)}$	erro (SDabs)	$^{87}Sr/$ $^{86}Sr_{(i)}$	Rb/Ba	$^{87}Rb/$ ^{86}Sr	erro (SDabs)	$^{147}Nd/$ ^{144}Nd	Erro ppm
*MH-15	1773.69	0.70755	0.00076	0.70716	0.05041	0.04859	0.04859	0.10617	30
*MH-13	1113.86	0.70765	0.00014	0.70656	0.07890	0.14012	0.14012	0.08029	74
*MH-14	1782.61	0.71012	0.00075	0.68567	0.07359	3.11922	3.11922	0.12468	28
*SSB-7	1642.67	0.71164	0.00026	0.70616	0.14778	0.69806	0.69806	0.13228	26
*SSB-32b	1643.06	0.71120	0.00360	0.71009	0.08169	0.14134	0.14134	0.12984	26

RODEIO VELHO MEMBER

Sample	SiO_2	Al_2O_3	Ba/Th	Sm (ppm)	Nd (ppm)	$^{143}Nd/$ ^{144}Nd	Erro ppm	$\varepsilon Nd_{(0)}$	$\varepsilon Nd_{(t)}$
*PH-	50.14	14.64	512.9	15.48	88.43	0.51165	14	-19.36	-13.92
*PH-4	65.53	12.7	1032.5	13.06	72.81	0.51168	14	-18.73	-13.45
*RLP-1	51.89	15.75	387.59	13.81	84.30	0.51191	14	-14.24	-8.39
*RLP-10	57.12	14.58	224.3	22.28	142.09	0.51185	13	-15.32	-9.21
*RPL-12	52.33	15.77	359.3	14.16	82.60	0.51170	13	-18.37	-12.80
*RLP-14	51.69	15.79	356.8	13.84	82.45	0.51173	16	-17.65	-11.95
*RLP-15	54.63	15.34	503.2	12.76	78.98	0.51174	9	-17.58	-11.65
*RLP-17	46.84	15.77	712.5	8.95	50.14		13	-17.58	-8.57

Sample	TDM	$^{87}Sr/$ $^{86}Sr_{(m)}$	erro (SD bs)	$^{87}Sr/$ $^{86}Sr_{(i)}$	Rb/Ba	87Rb/ 86Sr	erro SDabs	$^{147}Nd/$ ^{144}Nd	Erro ppm
*PH-3	1963.39	0.70579	0.00014	0.70555	0.00876	0.03662	0.00037	0.10581	28
*PH-4	1966.81	0.70750	0.00017	0.70665	0.00515	0.12658	0.00127	0.10848	28
*RLP-1	1500.53	0.70683	0.00021	0.70549	0.03558	0.20011	0.00200	0.09905	28
*RLP-10	1517.66	0.70788	0.00017	0.70774	0.00668	0.01975	0.00020	0.09481	26
*RPL-12	1854.20	0.70542	0.00017	0.70480	0.02521	0.092951	0.00093	0.10362	26
*RLP-14	1768.65	0.70526	0.00017	0.70467	0.02451	0.08756	0.00088	0.10149	32
*RLP-15	1705.90	0.71040	0.00016	0.70666	0.05833	0.55977	0.00560	0.09766	18
*RLP-17	1599.50	0.70739	0.00015	0.70685	0.02280	0.08146	0.00081	0.10791	26

Table 13. Geochemical analysis of the major and trace elements, Nd, Sr and Sm isotope results (modified from Almeida *et al* 2005). Major elements are in wt%; trace elements are in ppm. * = flows; ** = tuffs; # = welded tuffs. The analysis were performed at Activation Laboratories (ACTLAB), Canada, using the Argonium Plasma Spectrometry (ICP). The isotope analysis were carried out the Laboratory for Isotope Geology-LGI (Institute of Geosciences, Federal University of Rio Grande do Sul-UFRGS, Brazil).

All lavas, whichever unit they belong to, show a negative εNd in t or in the present time. The negative εNd values meaning in acid lavas is not always easily addressed. It could be related to crust derived lavas, basaltic lavas crustal contamination during differentiation, or mantle-derived, metasomatic lavas. However, negative εNd values in basalts are usually related to the latter two processes. Crustal contamination, either in acid or mafic lavas generally also affects the Rb/Sr system, increasing the radiogenic Sr amount and conferring to the rocks, a high initial $^{87}Sr/^{86}Sr$ ratio. Acampamento Velho Formation and Rodeio Velho Member mafic lavas display εNd ranging from -2.9 to -10.3 and from -8.4 to -13.9, respectively, and an initial $^{87}Sr/^{86}Sr$ ratio from 0.706 to 0.707 and from 0.704 to 0.707. These initial $^{87}Sr/^{86}Sr$ ratios are consistent with rocks derived from a Neoproterozoic depleted mantle, represented by the Cambaí Complex rocks (Babinski et al., 1996), plotted in figure 26. The negative correlation between initial $^{87}Sr/^{86}Sr$ ratio and SiO_2 content observed in figure 27a, along with the constancy of the SiO_2 content, despite εNd values (Fig. 27b) for Acampamento Velho Formation and Rodeio Velho Member mafic lavas ratify the idea that the crustal contamination was not significant during the differentiation of these rocks, in the early stages of differentiation (at least until these rocks have reached around 55 wt% of SiO_2 content). Crustal contamination or crustal origin is assumed for rocks with SiO_2 content higher than 55 wt%. It can also explain the occurrence of acid lavas and pyroclastic material in the Acampamento Velho Formation (Fig. 27a and b). A plot of initial $^{87}Sr/^{86}Sr$ against Ba/Th ratios (Fig. 28) shows strong enrichment in radiogenic Sr and Th in the acid rocks, which could be related to a crustal component. In order to better constrain the possible sources of the Acampamento Velho Formation and Rodeio Velho Member lavas, juvenile Neoproterozoic depleted mantle derived rocks (the Cambaí Complex) and Neoproterozoic crustal rocks (arkosic quartzites, taken as representative of the average composition of the crust and meta-pelagic sediments) were plotted with Acampamento Velho Formation and Rodeio Velho Member samples (Fig. 26). The isotopic ratios and compositions of these rocks were calculated for 550 Ma and mixture lines were calculated based on two-end-member models.

From the analysis of figure 26, the Acampamento Velho Formation mafic lavas could be regarded as a mixture of depleted mantle-derived basalt plus 20% to 30% of crustal contamination. The crustal material matches best Neoproterozoic arkosic quartzites rather than pelagic sediments. Acampamento Velho Formation felsic lavas evolution trend displays an increase in crustal contamination. However, Rodeio Velho Member mafic lavas can not be easily explained by crustal contamination of depleted mantle derived magma. The composition of these lavas request an end member highly enriched in radiogenic Nd, but impoverished in radiogenic Sr. An enriched mantle type I (EM I according to Zindler & Hart, 1986) is such a reservoir, well established for Phanerozoic rocks with the same trend as Rodeio Velho Member (Fig. 26).

The differences between Acampamento Velho Formation and Rodeio Velho Member mafic samples can be highlighted using Ba/Th ratio (Fig. 28). Both mafic lavas show preferential enrichment in Ba relative to Th. However, the mafic lavas of the RVM show distinctive enrichment in the Ba/Th ratio (due to its very high Ba content, up to 2695 ppm) without a change in the $^{87}Sr/^{86}Sr$ initial ratio. On the other hand, Acampamento Velho Formation mafic lavas show a weak enrichment in Ba/Th and $^{87}Sr/^{86}Sr$ initial ratios. Such behavior suggests that Rodeio Velho Member was originated from a depleted mantle, re-enriched in highly incompatible elements such as Ba.

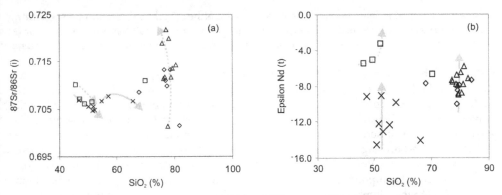

Fig. 27. (a) $^{87}Sr/^{86}Sr$ initial versus SiO_2 and (b) $\varepsilon Nd(t)$ versus SiO_2 diagrams for the Acampamento Velho Formation and Rodeio Velho Member volcanic rocks. Symbols are the same as in figure 25 (Almeida et al., 2005).

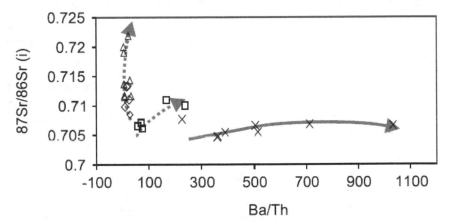

Fig. 28. Initial $^{87}Sr/^{86}Sr$ versus Ba/Th diagram for the isotope data obtained for the volcanic rocks from Acampamento Velho Formation and Rodeio Velho Member. Observe the different trends for these groups as well as the radiogenic Sr and Th enrichment in the Upper Felsic Association, and in a lesser degree for the mafic rocks of the Lower Mafic Association from Acampamento Velho Formation. Symbols are the same as in figure 25. (Almeida et al., 2005).

Regarding the Nd model ages (Table 13), we can see Acampamento Velho Formation and Rodeio Velho Member mafic associations with values raging from 1.1 to 1.8 Ga and 1.5 to 2.0 Ga, respectively. These model ages point to also for a modified mantle at end of Brasiliano Cycle, in post-orogenic tectonic setting (during the Upper Neoproterozoic to Eopaleozoic). However, Acampamento Velho Formation has some juvenile component with already modified mantle since it displays somewhat younger model ages and lesser negative εNd values. On other side, the Acampamento Velho Formation felsic association Nd model age is from 1.3 to 2.8 Ga (Table 13). This information coupled with Acampamento Velho Formation felsic association lower εNd negative values, if compared to those of mafic association, suggest at least two alternatives, namely, either (1) the felsic magma is a melting of a lower crust with a dominant Paleoproterozoic age or mixed Brasiliano-older crust, or (2) the felsic magma may be derived from the mixture of mafic magma and a crustal enrichment component as described below.

In this case, the existence of the different sources in the mantle during the Neoproterozoic magmatism in the Camaquã Basin could be connected with a sedimentary component. The subduction process could be responsible for 20-30% of contamination by sediments, probably Neoproterozoic arkosic quartzites, and the establishment of a reservoir. The generated magma at some point initiated its ascension under conditions of high oxygen fugacity. At a certain stage, this magma separated into two fractions: a mafic portion, from which Acampamento Velho Formation mafic lavas were originated, and another one, enriched in crustal component, which gave rise to the AVF felsic rocks. The rocks belonging to the Rodeio Velho Member seem to have been originated from a different reservoir, much more enriched in incompatible elements and depleted in radiogenic Sr with EMI I characteristics.

7. Camaquã Basin Volcanism-tectonic environment - Conclusions

The Camaquã Basin developed during the final stages of the Brazilian-Pan-African Cycle (700 Ma – 500 Ma.) in the Dom Feliciano Belt. This basin was formed in a retroarc setting and was filled essentially by clastic sediments and volcanogenic rocks (Chemale, 2000). The Maricá Formation (Leinz et al., 1941) is the lowest sedimentary sequence for Camaquã Basin formed as retroarc foreland basin, and comprises a fluvial to marine sedimentation formed due to the load of the adjacent granitic-gneissic Pelotas Belt and volcano sedimentary Porongos Belt (Gresse et al., 1996; Paim et al., 2000), which contains depositional sequences formed in fluvial and marine sediments. The deposition interval of the Maricá Formation is between 601 and 592 Ma based on the age obtained of detrital zircon from Maricá Formation sandstone, at 601±13 Ma, and the age of the Hilário Member, which overlies in unconformity the Maricá Formation units, at 590 ± 5.7 Ma (Janikian et al., 2008) and 591±3.0 Ma (Fig. 29). The geochemical and isotope signature of the Maricá Formation sediments point to reworked crustal material based on the Sm-Nd data (Borba et al., 2008). The detrital zircons of the sample CK 239D have contribution of the Brasiliano magmatic arc material and also some Paleoproterozoic crustal material (Siderian, Rhyacian and Statherian) and Neoarchean.

The overlying Bom Jardim units consist of a typical continental sedimentation with fluvial, delta and lacustrine sediments with a strong presence of intermediate shoshonitic magmatism at the base, represented by Hilário Andesites and associated to lamprophyres (Lima et al., 1995). These magmatic rocks are very well constrained by U-Pb ages obtained for Hilario Formation at 590 ± 5.7 Ma (Janikian et al., 2008) and 591.8 ±3.0 Ma (this work) (Fig. 29). The last age was obtained by us using a lamprophyric sample (alkaline basalt). Considering that the lamprophyres correspond to the last manifestation of the Hilário volcanism, we can suggest that this event is positioned roughly between 592 and 590 Ma. The volcanic rocks that compose the Bom Jardim Group base are related to the calc-alkaline magmatism from shoshonitic affinity (Nardi & Lima, 1985; Lima et al., 1995) and also associated with a stabilization phase of the plate subduction process (Lima & Nardi, 1992; Chemale et al., 1999; Almeida et al., 1999). The tectonic environment of the Bom Jardim Group units is thus related to the late to post orogenic processes of Dom Feliciano Belt, from 591.8 Ma to < 573 Ma, during which, the sedimentation of the Camaquã Basin was dominated by strike-slip tectonics with some transtensional/transpressional components in retroarc position. The Bom Jardim Group is limited by two angular discordances that are confined to the Maricá Group (base) and Acampamento Velho Formation volcanic rokcs (top). The Acampamento Velho is a bi-modal magmatic association with some sedimentary contribution (as Santa Fé Member of Paim et al., 2000) and is the base of a volcano-sedimentary sequence, the Acampamento Velho/Santa Bárbara Group. This sequence begins with basic and acid volcanic magmatism of the alkaline signature, which is overlied by the three Santa Bárbara Group sedimentary sequences. The lower and intermediate sedimentary sequences (Santa Fé and Lanceiros Formation) represent a fluvial-deltaic sedimentation, and the upper formation (Pedra do Segredo Formation), overlies in angular unconformity on Lanceiros Formation. Units are composed by a coarsening up alluvial-fluvial depositional sequence.

The U-Pb zircon data suggest that Acampamento Velho Rhyolites formed from 570 to 544 Ma, tanking account the SHRIMP age obtained by Sommer (2005) for Ramada Plateau acid volcanic rocks of the is 549 ± 5 Ma and by Janikian et al. (2008) for acid volcanic rocks in the Caçapava region is 574±7 Ma. However, we obtained the U-Pb zircon age of 553 ± 5 Ma for Acampamento Velho Formation using an andesitic basalt sample of the lower mafic association, which is close to that age of Ramada Plateau. In the Santa Bárbara (Fig.29, west column) and Guaritas sub-basins (Fig. 29, east column), the basic flows from basal contact, pyroclastic manifestations and rhyolitic flows (from the base to the top) of the Acampamento Velho are in unconformity with the Bom Jardim Group, and the upper contact is made by unconformity with the Sequence I (or Santa Fé and Lanceiros Formation base from Santa Bárbara Group). The observed stratigraphic position in field – basalts and andesitic basalts on the base and felsic sequence on the top (Figs. 12 and 29), as well as the geochemical and isotopic behavior, show that the first phase formed the basalts and andesitic basalts, and later, the rhyolitic rocks, which confirm that the Acampamento Velho Formation is bimodal volcanism. The best estimated ages for the Acampamento Velho Formation and Santa Barbara Group are 553 and 549 Ma, during which, an extensional events controlled the deposition and magmatism.

The Santa Barbara Group units are overlied in angular unconformity by the sediments of the Guaritas Group. At the base of this group, occurs the alkaline basalt lavas called as Rodeio Member. We obtained the age of 547 ± 6.3 Ma for the Rodeio Velho Member alkaline basalt, which can be interpreted the maximum age for the magmatism, therefore, the deposition of the Guaritas Formation. On other side, detrital zircon collected from the upper portion of the Guaritas Group is dated at 535±10 Ma (Hartmann et al. 2008). The present data suggest a Cambrian age for the Rodeio Velho Formation, instead the U-Pb age of 470 ± 19 Ma (sensitive high mass-resolution ion microprobe - SHRIMP) obtained by Hartmann et al. (1998) for the Rodeio Velho Member. Santa Bárbara Sub-basin (Fig. 29 - west column), which is above the Rodeio Velho Member (Pedra Pintada Formation), is in contact by disconformity with the Sequence III (or Segredo Formation). Guaritas Sub-basin (Fig. 29 - east column), that is above the Rodeio Velho (and Pedra Pintada Formation), is in contact by disconformity and we can see the Santa Bárbara Group sedimentary rocks (Fambrini, 2005). Thus, beyond the U-Pb ages, the stratigraphic position showing the Rodeio Velho Member above Acampamento Velho Formation, allow us to postulate that first was generated the Acampamento Velho volcanism, and then, the Rodeio Velho volcanism. The Guaritas Group sedimentation occurred under more stable conditions in the retroarc region, and it is deposited under a transtensional tectonics.

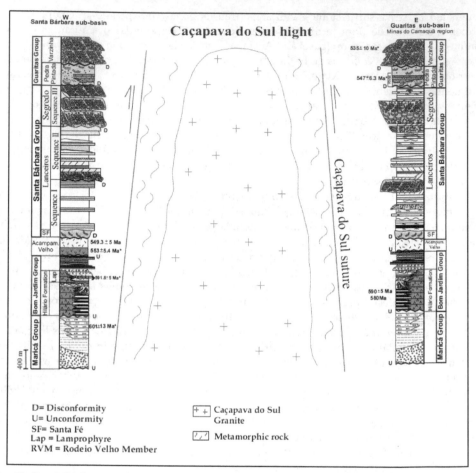

Fig. 29. Santa Bárbara and Guaritas sub-basin: stratigraphic columns including the main unconformity and disconformity, which bounds the Camaquã Basin unit. West column is modified from Paim et al (1995, 2000), Almeida et al (2002) and Borba & Misuzaki (2003). East column is modified from Paim et al. (1995), Wildner (2002), Fambrini et al. (2005) and Janikian et al. (2003, 2008).

The Rodeio Velho Member is interbedded or arranged in volcano-sedimentary features of the interaction (flow striation, xenoliths, clastic dykes, marks in increasing and peperites) with the Pedra Pintada Formation sedimentary rocks from Guaritas and Santa Bárbara sub-basins. We would like to point out that, between these two sub-basins, there is the Caçapava structural hill, the Caçapava Granite (average U-Pb age of 550 Ma) and the Caçapava do Sul Suture (Fig. 1 and Fig. 29). According to Nardi and Bitencour (1989), the Caçapava Granite was positioned in mesozone level, with microgranitiques and aplitiques rocks positioned at summit of the body and mainly associated to leucogranitiques. Considering these information, we may suggest that the granite together with the rocks above (more than 6 km thick) may have suffered uplift and denudation processes during a long time before the granite emerge. These same authors mentioned that the Caçapava do Sul Granite Complex is a body composed of two contrasting blocks bounded by NW-trending structures. Borba et al. (2002) from the study about apatite fission-track (FT) thermochronology for samples of the Caçapava do Sul granites determined to the Northern block, an apparent FT ages of 293.5 and 274.8 Ma, and an estimated for the initial track recording an age of 366 Ma. These results are interpreted in the frame of the uplift and denudation history of the region during the Paleozoic-Mesozoic intraplate Parana´ Basin evolution, as influenced by the convergent tectonic setting of Southwestern Gondwana. In the Southern block, where dioritic and gneissic portions crop out and apparent FT ages range from 252.1 to 245.5 Ma. A possible subsequent Upper Cretaceous uplift event can be inferred from the modeled thermal history curves and the apparent FT age of the Northernmost collected sample is 73.7 Ma. If it is considered the presence of aplites and microgranites at the summit of the body, than it can be said that it has suffered little denudation after have emerged. Costa (1997), based on geophysical data interpretations, mentioned that Caçapava do Sul Suture corresponds to a deep geophysical anomaly that separates two magnetic domains. Therefore, it is important to consider an independent basin evolution for each suture side and beyond that this sub-basin may have shown different subsidence degree. If we consider that Guaritas Sub-basin the Santa Bárbara Group rocks show marked dips, and the Guaritas Group rocks do not show this feature, while that in Santa Bárbara sub-basin, the rocks Guaritas Group´s base (Pedra Pintada Formation) according to Almeida et al. (2003a), the layers exhibit latitudes 188°/40 ° NW, 194° /23 ° NW e 195° /26° NW, then, we could suggest that tilting would be produced by biggest Caçapava Suture influence (when reactivated), which is in direct contact with these rocks. Paim et al. (2000) attributed tilting of these rocks as a reflex of normal falls, and subordinately strike slip movements. Thus, these declivities can be formed due to basin different subsidences.

Almeida et al. (2002, 2003a, 2003b and 2005) based on Brazilian Orogeny evolution pattern from Chemale Jr. (2000); considered that the Acampamento Velho Formation was generated in an extensional regime preceding the Rio de la Plata and Kalahari continental plate collisions. The Rodeio Velho Member may correspond to the late magmatic manifestation and may have occurred after the collisional phase between Rio de la Plata (with Encantadas Microcontinent attached) and Kalahari Continental Plate. In the context of adopted pattern, it would correspond to the Phase IV of the same author (Fig. 2 and Fig. 30). Thus, we propose that between 560-550 Ma, considering only the Camaquã Basin central portion, it may exist two extensional basins depocentre, such as Santa Bárbara and Guaritas sub-basins (Fig. 30b), being them limited by Caçapava Hill. These sub-basins were already partially filled by coastal deposits at their base, including fluvial and shallow marine sediments, with scarce volcanic contribution and U-Pb age of 601 Ma (Maricá Group). On these rocks, is an unconformity named as Bom Jardim Group, composed of volcanic rocks at the base (Hilário Formation) and sediments at the top. On these rocks, in disconformity, are volcanic rocks of the basaltic and rhyolitic composition (Acampamento Velho). According to Almeida et al (2005, 2007), the generated magma (of Acampamento Velho) at some point initiated its ascension. At a certain stage, this magma separated into two fractions: a mafic portion and another one, enriched in crustal component. Between 560-550 Ma, after the magmatic chamber formation as well as the separation of the magma into two fractions, both fraction magmas rise and mafic flow manifestations (lower mafic association) took place around 553 Ma.

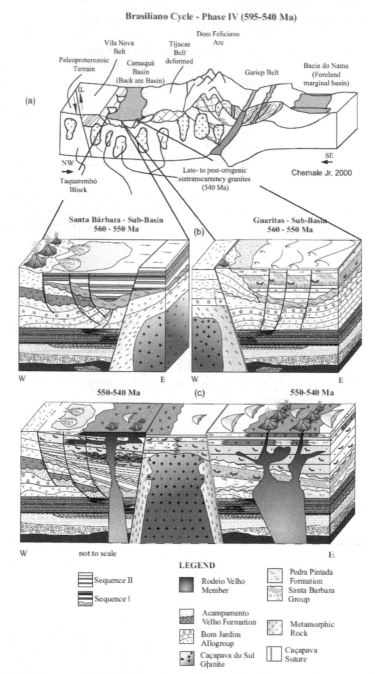

Fig. 30. Schematic diagram block showing the position of the different rock units in the: (a) Santa Bárbara and (b) Guaritas sub-basins between 560-550 Ma and 550-540 Ma. These sub-basins were generated during the Phase IV of Brasiliana Orogeny established by Chemale Jr. (2000).

The other magma fraction underwent through a significant enrichment in crustal component. This viscous magma slowly rise around 549 Ma revealing itself in a first phase as pyroclastic rhyolitic rock (tuff, lapillitic tuff and welded tuff) and afterwards as a rhyolitic flow (the upper felsic association). In the range of 547-535 Ma (Fig. 30c), an alkaline mantelic magma may ascend in the last magmatism stages by the deep faults during an intra-plate mechanic extensional event, generating the Rodeio Velho Member event (Fig.30c). The Rodeio Velho Member volcanic manifestations took place as flows interbedded which sometimes show interactions of the lava with sediments poorly consolidated, and correspond to the Pedra Pintada Formation eolian deposits from Guaritas Group. We can observe these features at Santa Bárbara and Guaritas sub-basins.

8. Acknowledgements

Delia del Pilar M. de Almeida gratefully acknowledges the fellowship (151477/2007-8 process) granted by Conselho Nacional de Desenvolvimento Científico e Tecnológico-CNPq (Brazil), which made this work possible and feasible. The authors are grateful to the Universidade Federal do Pampa (UNIPAMPA) for the scholarship granted to the undergraduate students Taís Regina Cordeiro de Oliveira and Igo Silva de Almeida that collaborated in the figures edition and text formatting.

9. References

Almeida, F. (1969). Diferenciação Tectônica da Plataforma Brasileira. In: *Congr. Bras. Geol., 1*, Salvador, pp. 29-46.

Almeida, F.; Hasui,Y. & Brito Neves, B. (1976). The Upper Precambrian of South America. *Boletim IG/USP*, Vol.7, pp.45-80.

Almeida, F.; Hasui,Y.; Brito Neves, B. & Fuck R. (1981). Brazilian strutural provinces: an introduction. *Earth Science Ver*, Vol.17, pp.1-29.

Almeida, D. del P.; Pereira, V.; Machado, A.; Zerfass, H. & Freitas, R. (2007). Late Sodic Metasomatism evidences in bimodal Volcanic Rocks of the Acampamento Velho Alloformation, Neoproterozoic III, Southern Brazil. *Anais da Academia Brasileira de Ciências*, Vol.79, Nº4, pp.1-13, ISSN 001-3765.

Almeida, D.del P.; Conceição, R.; Chemale Jr. F.; Koester, E.; Borba, A. & Petry, K. (2005). Evolution of heterogeneous Mantle in the Acampamento Velho and Rodeio Velho volcanic events, Camaquã Basin, Southern Brazil. *Gondwana Research*, Vol.8, Nº4, pp.479-492, ISSN 1342-937X.

Almeida, D.del P.; Zerfass, H.; Basei, M. & Lopes, R. C. (2003a). Eventos vulcânicos alcalinos na Bacia do Camaquã: o vulcanismo Neoproterozoico III Acampamento Velho e o magmatismo Meso-Ordoviciano (?) Rodeio Velho. In: *Caracterização e Modelamento de Depósitos Minerais*, Ronchi, L.H., Althoff, F., (Eds.), 325-350, Universidade do Vale do Rio dos Sinos, São Leopoldo, Brazil.

Almeida, D. del P.; Hansen, M; Fernsterseifer H.; Petry K. & Lima L.. (2003b). Petrology of a subduction-related caldera and post-collisional, extesion-related volcanic cones from the Early Cambrian and Middle Ordovician (?) of the Camaquã Basin, southern Brazil. *Gondwana Research*, Vol.6, Nº3, pp.541-552, ISSN 1342-937X.

Almeida, D. del P.; Zerfass, H.; Basei, M.; Petry, K. & Gomes, C. (2002). The Acampamento Velho Formation, a Lower Cambrian bimodal volcanic package: geochemical and stratigraphic studies from the Cerro do Bugio, Perau and Serra de Santa Bárbara (Caçapava do Sul, Rio Grande do Sul, RS – Brazil). *Gondwana Research*, Vol.5, No3, pp. 721-733, ISSN 1342-937X.

Almeida, D. del P.; Lopes R. C.; Lima L. & Gomes, C. (2000). Petrography and Geochemistry of the Volcanic Rocks of the Rodeio Velho Member, Ordovician of the Camaquã Basin (Rs-Brazil): Preliminary Results. *Revista Brasileira de Geociências*, Vol.30, No4, pp.23-34, ISSN 0375-7536.

Almeida, D. del P.; Lima L. & Gomes, C. (1999). Região de Vista Alegre-Lavras do Sul/RS: Uma Zona Tipo da Formação Hilário. *1 Simposio sobre Vulcanismo e Ambientes Associados, Gramado – RS*. Vol.1,pp.56.

Babinski, M.; Chemale Jr. F.; Hartmann, L.; Van Schmus, W. & Silva, L. (1996). Juvenile accretion at 750-700 Ma in Southern Brazil, *Geology*, Vol.24, No5, ISSN 0091-7613.

Basei M.; Siga J.; Masquelin H.; Harara M.; Reis Neto, J. & Preciozzi P. (2000). The Dom Feliciano Belt of Brazil and Uruguay and Its Foreland Domain, the Rio de La Plata Craton. *Tectonic Evolution of South America*, Cordani, U.G. et al. (Eds.), pp. 311-334, Rio de Janeiro, Brazil.

Borba, A.; Mizusaki, A. M.; Santos, J.; McNaughton, N.; Onoe, A. & Hartmann, L. (2008). U-Pb zircon and [40]Ar-[39]Ar K-feldspar dating of syn-sedimentary volcanism of the Neoproterozoic Maricá Formation: constraining the age of foreland basin inception and inversion in the Camaquã Basin of southern Brazil. *Basin Research*, Vol. 20, pp 359-375, ISSN 0950-091X.

Borba, A., Maraschin, A. & Mizusaki, A. M. (2004). Stratigraphic analysis and depositional evolution of the Neoproterozoic Marica Formation (southern Brazil): constraints from field data and sandstone petrography. *Gondwana Res.*, Vol. 7 No 3, pp. 871-886, ISSN 1342-937X.

Borba, A. & Mizusaki, A. M. (2003). Santa Bárbara Formation (Caçapava do Sul, southern Brazil): depositional sequences and evolution of an Early Paleozoic post-collisional basin. *Journal of South America Earth Sciences*, Vol.16, No5, pp.365-380, ISSN 0895-9811.

Borba, A., Vignol-Lelarge, M.L., Mizusaki, A. M. (2002). Uplift and denudation of the Cac‚apava do Sul granitoids (southern Brazil) during Late Paleozoic and Mesozoic: constraints from apatite fission-track data. *Journal of South American Earth Sciences*, Vol. 15, pp. 683–692, ISSN 0895-9811.

Bonhomme, M. & Ribeiro, M. (1983). Datações K-Ar das argilas associadas a mineralização de cobre da Mina Camaquã e de suas encaixants. In: *1 Simpósio Sul-Brasileiro de Geologia*, Vol. 1, pp. 82-88.

Chemale Jr., F. (2000). Evolução geológica do Escudo Sul-rio-grandense. *Geologia do Rio Grande do Sul*, Holz, M. & De Ros, L.F. (Eds.), pp. 13-52, Universidade Federal do Rio Grande do Sul, Porto Alegre, Brazil.

Cordani, U.; Halpern, M. & Berenholc, M. (1974). Comentários sobre as determinações da Folha de Porto Alegre. In: DNPM, Carta Geológica do Brasil ao Milionésimo, texto explicativo da Folha de Porto Alegre e Lagoa Mirim, pp.70-84.

Costa A. (1997). Teste e modelagem geofísica da estruturação das associações litotectônicas Précambrianas no Escudo Sul-rio-grandense. Tese de Doutorado, Instituto de Geociências, Universidade Federal do Rio Grande do Sul Brazil, 291 pp.

Cullers, R. & Graf, J. (1984). Rare earth elements in igneous rocks of the continental crust: predominantly basic and ultrabasic rocks. In: *Rare earth Element Geochemistry*, P. Anderson (ed.), 237-274, Amsterdam.

Fambrini G.; Janikian L; Paes de Almeida, R. & Fragoso-César, A. R. (2005). O Grupo Santa Bárbara (Ediacarano) na Sub-Bacia Camaquã Central, RS: estratigrafia e sistemas deposicionais. *Revista Brasileira de Geociências*, Vol.35, Nº2, pp.227-238, ISSN 0375-7536.

Fragoso-Cesar, A. R.; Fambrini, G.; Almeida, R.; Pelosi, A. P.; Janikian, L.; Nogueira, A. & Riccomini, C. (2001). As Coberturas do Escudo Gaúcho no Rio Grande do Sul: Revisão e Síntese (Abst.). *XI Congreso Latinoamericano de Geología e III Congreso Uruguayo de Geología*, Uruguay.

Fragoso-César, A. R.; Fambrini, G.; Paes de Almeida, R.; Pelosi, A.; Janikian, L.; Riccomini, C.; Machado, R.; Nogueira, A. & Saes G. (2000). The Camaquã extensional basin:Neoproterozoic to early Cambrian sequences in southernmost Brazil, *Revista Brasileira de Geociências*, Vol.30, pp.438-441, ISSN 0375-7536.

Fragoso-Cesar, A. R., Faccini, U., Paim, P., Lavina, E. & Flores, J. (1985). Revisão na estratigrafia das molassas do Ciclo Brasiliano no Rio Grande do Sul. *2 Simpósio Sulbrasil. Geol.*, pp. 477-491.

Fragoso-Cesar, A. R., Lavina, E., Paim, P. & Faccini,U. (1984) A antefossa molássica do cinturão Dom Feliciano no Escudo do Rio Grande do Sul. *33 Congresso Brasileiro de Geologia*, pp. 3272-3283.

Fragoso-Cesar, A. R. (1982). Associações petrotectônicas do Cinturão Dom Feliciano (SE da Plataforma Sul-Americana) (ext. abst.), *1 Congresso Brasileiro de Geologia*, Vol. 1, pp.1-12, Salvador, Brazil.

Goñi, J.; Goso, H. & Issler R. (1962). Estratigrafía e Geologia Económica do Pré-Cambriano e Eo-Paleozóico uruguaio e sul-Riograndense. *Avulso da Escola de Geologia*, UFRGS, Vol.3, pp.1-105.

Gresse, P.; Chemale Jr., F.; Silva, L.; Walraven, F. & Hartmann, L. (1996) Late- to post-orogenic basins of the Pan-African- Brasiliano collision orogen in southern Africa and southern Brazil. *Basin Research*, Vo.8, pp.157-171, ISSN 0950-091X.

Hartmann, L.; Santos, J. & McNaughton, N. (2008). Detrital zircon U-Pb age data, and Precambrian provenance of the Paleozoic Guaritas Formation, Southern Brazilian Shield. *International Geology Review*, Vol. 50, pp. 364–374, ISSN 0020-6814.

Hartmann, L.; Silva, L.; Remus, M.; Leite, A. & Philipp R. (1998). Evolução Geotectônica do sul do Brasil e Uruguai entre 3,3 Ga e 470 Ma. *2 Congreso Uruguayo de Geologia*, pp.277-284.

Horbach, R.; Kuck, L; Marimon, R.; Moreira, H. L.; Fuck, G.; Moreira, M.; Marimon, M.; Pires, J.; Vivian, O.; Marinho, D. & Teixeira, W. (1986). Geologia Folha Sh. 22 (Porto Alegre) e parte das folhas Sh. 21 (Uruguaiana E Si. 22 (Lagoa Mirim). Vol. 33, pp. 29-312, Rio de Janeiro, IBGE.

Janikian, L.; Almeida, R.; Ferreira da T.; Fragoso-Cesar, A. R.; Souza D'A., M.; Dantas, E. & Tohver, E. (2008). The continental Record of Ediacaran volcano-sedimentary successions in southern Brazil and their global implications. *Terra Nova*, Vol. 20, pp. 259-266, Publ. Online – doi: 10.1111/j.1365-3121.2008.00814.x, ISSN 0954-4879.

Janikian, L.; Almeida, R.; Fragoso-Cesar, A. R. & Fambrini, G. (2003). Redefinição do Grupo Bom Jardim (Neoproterozóico III) rm sua área tipo: litoestratigrafia, evolução paleoambiental e contexto tectônivo. *Revista Brasileira de Geociências*, Vol.33, No 4, pp. 349-362, ISNN 0375-7536.

Leinz, V.; Barbosa, A. F. & Teixeira, E. (1941). Mapa Geológico Caçapava-Lavras. Boletim 90, Secretaria da Agricultura, Indústria e Comércio, RS.

Leite, J.; Hartmann, L.; McNaughton, N. & Chemale Jr., F. (1998). SHRIMP U/Pb zircon geochronology of Neoproterozoic juvenile and crustal-reworked terranes in southernmost Brazil. *International Geology Review*, Vol. 40, No 8, pp. 688-705, ISNN 0020-6814.

Leite, J. (1997). A origem dos Harzburgitos da Sequência Cerro Mantiqueiras e implicações tectônicas para o desenvolvimento do Neoproterozóico no Sul do Brasil. Unpublished, PhD Thesis, Universidade Federal do Rio Grande do Sul, 243 p.

Leite, J.; McNaughton, N.; Hartmann, L. & Chemale Jr., F. (1995). Age and tectonic setting of metabasalts and metagranitoids from the Cerro Mantiqueiras region: evidences from SHRIMP U/Pb zircon dating and Pb/Pb. In: *5 Simpósio Nacional de Estudos Tectônicos*, Vol. 1, pp. 389-390.

Leites, S.; Lopes, R. C.; Wildner, W.; Porcher, C. & Sander, A. (1990). Divisão litofaciológica da Bacia do Camaquã na folha Passo do Salsinho, Caçapava do Sul, RS, e sua interpretação paleoambiental. *36 Congresso Brasileiro de Geologia*, pp.300-312.

Lima, L.; Almeida, D. del P.; Collao, S. (2001). El distrito Minero de Camaquã: Um ejemplo de mineralizaciones tipo epi-mesotermales alojadas em rocas sedimentarias. *Acta Geologica Leopoldensia*, Vol.23 No 51, pp.85-102, ISNN 0102-1249

Lima, E. & Nardi, L. (1998). The Lavras do Sul Shoshonitic Association: implications for origin and evolution of Neoproterozoic Shoshonitic magmatism in southermost Brazil, *Journal of South American Earth Sciences*, Vol.11, pp.67-77, ISNN 0895-9811.

Lima, E.; Wildner W.; Lopes R.C., Sander, A. & Sommer C. (1995). Vulcanismo Neoproterozóico associado às bacias do Camaquã e Santa Bárbara – RS: Uma revisão, *VI Simp. Sul-Brasileiro de Geologia / I Encontro de Geologia do Cone Sul*, pp. 197-199.

Lopes, R. C.; Wildner, W.; Sander, A. & Camozzato, E. (1999). Alogrupo Guaritas: aspectos gerais e considerações sobre o posicionamento do vulcanismo Rodeio Velho (encerramento do Ciclo Brasiliano ou instalação da Bacia do Paraná) (Abst.), *1 Simpósio Sobre Vulcanismo e Ambientes Associados*, pp. 17, Gramado, Brazil.

Nardi, L. & Lima, E. (1985). A associação shoshonítica de Lavras do Sul, RS. *Revista Brasileira de Geociências*, Vol.15, pp.139-146, ISNN 0375-7536.

Nakamura, N. (1977). Determination of REE, Ba, Fe, Mg, Na and K in carbonaceous and ordinary chodrites. *Acta Geochimica et Cosmochimica*, Vol.38, pp.757-775.

Paim, P.; Chemale Jr., F. & Lopes, R. C. (2000). A Bacia do Camaquã. *Geologia do Rio Grande do Sul*, Holz, M., De Ros, L.F. (Eds.), 231-274, Universidade Federal do Rio Grande do Sul, Porto Alegre, Brazil.

Pearce, J. & Norry, M. (1979). Petrogenetic Implications of Ti, Zr, Y, and Nb Variations in Volcanic Rocks. *Contributions to Mineral and Petrology*, Vol.69, pp.3-47, ISSN 0010-7999.

Petry, K. (2006). Feições de interação volcano-sedimentares: seu uso como indicadores de contemporaneidade no Magmatismo rodeio Velho (Meso-Ordoviciano) e no vulcanismo Serra Geral (Cretáceo inferior), Unpublished, Dissertação de Mestrado, Universidade do Vale do Rio dos Sinos (UNISINOS), 91pp.

Remus, M.; McNaughton, M.; Hartmann, L. & Fletcher, I. (1999). Gold in the Neoproterozoic juvenile Bossoroca Volcanic Arc of southermost Brazil: isotopic constrains on timing and sources, *Journal of South America Earth Sciences*, Vol. I2, pp. 349-366, ISNN 0895-9811.

Remus, M.; McNaughton, N.; Hartmann, L. & Fletcher, I. (1997). Zircon SHRIMP dating and Nd isotope data of granitoids of the São Gabriel Block, southern Brazil: evidence for an Archean Paleoproterozoic basement. II International Symposium on Granites and Associated Mineralization (pp. 271-272). II ISGAM, Salvador, BA, Brazil, SME-BA.

Remus, M.; McNaughton, N.; Hartmann, L. & Groves, D. (1996). SHRIMP U/Pb zircon dating at 2448 Ma of the oldest igneous rock in southern Brazil: identification of the westernmost border of the Dom Feliciano Belt. 1 *Symposium on Archean Terranes of the South American Platform*, Vol, 1, pp. 67-70.

Ribeiro, M. & Fantinel, L. (1978). Associações petrotectônicas do Escudo Sul-rio-grandense: I-Tabulação e distribuição das associações petrotectônicas do Rio Grande do Sul, *Inheríngia, Série Geológica*, Vol.5, pp.19-54, ISSN 0073-4705.

Ribeiro, M. & Teixeira, C. (1970). Datações de rochas do Rio Grande do Sul e sua influência nos conceitos estratigráficos e geotectônicos locais. *Inheríngia, Série Geológica*, Vol. 3, pp. 109-120, ISSN 0073-4705.

Ribeiro, M.; Bocchi; P.; Figueiredo Filho, P. & Tessari, R. (1966). Geologia da Quadrícula de Caçapava do Sul, Rio Grande do Sul – Brasil. Divisão de Fomento da Produção Mineral, Boletim 127.

Robertson, J. (1966). Revision of Stratigraphy and nomenclature of rock units in Caçapava-Lavras Region. IG-UFRGS, *Notas e Estudos*, Vol.1, N°2, pp.41-54, Porto Alegre, Brazil.

Sartori, P. (1978). Petrologia do Complexo Granítico São Sepé, RS. Ph.D. thesis, São Paulo University, 195 p.

Sartori, P. & Kawashita, K. (1985). Petrologia e geocronologia do Batólito Granítico de Caçapava do Sul, RS. 2 *Simpósio Sul-Brasileiro de Geologia*, Vol. 1, pp. 102-115.

Silva Fº, W.; Fragoso-Cesar, A. R.; Machado, R.; Sayeg, H.; Fambrini, G. & Ribeiro de Almeida, T. (1996). O magmatismo Rodeio velho e a Formação Guaritas no eopaleozóico do Rio Grande do Sul: Uma revisão (ext. abst.). 39 *Congresso Brasileiro de Geologia*, Vol. 5, pp. 433-435, Salvador, Brazil.

Simon, E; Jackson, S.; Pearson A.; Griffina W.; Belousova E. (2004). The application of laser ablation-inductively coupled plasma-mass spectrometry to in situ U–Pb zircon geochronology. *Chemical Geology*, Vol. 211, pp. 47-69, ISSN 0009-2541.

Soliani Jr., E.; Koester, E. & Fernandes, L. (2000). Geologia isotópica do Escudo Sul-rio-grandense, parte II: os dados isotópicos e interpretações petrogenéticas. *Geologia do Rio Grande do Sul*, Holz, M. & De Ros, L.F. (Eds.), 175-230, Universidade Federal do Rio Grande do Sul, Porto Alegre, Brazil.

Soliani Jr., E.; Fragoso-Cesar, A. R.; Teixeira, W. & Kawashita, K. (1984). Panorama geocronológico da porção meridional do Escudo Atlântico. 33 *Congresso Brasileiro de Geologia*, Vol. 5, pp. 2435-2449.

Sommer, C.; Lima, E.; Nardi, L.; Figueiredo, A. & Pierosan, R. (2005). Potassic and low- and high-Ti mildly alkaline volcanism in the Neoproterozoic Ramada Plateau, southermost Brazil. *Journal of South American Earth Sciences*, Vol.18, N°3, pp.237-254, ISSN 0895-9811.

Sommer, C.; Lima, E.; Nardi, L. (2000) . Evolução Do Vulcanismo Alcalino da Porção Sul do Platô do Taquarembó, Dom Pedrito, RS. *Revista Brasileira de Geociências*, Vol. 29, N° 2, pp. 245-254, ISSN 0375-7536.

Stacey, J.S., Kramers, J.D., 1975. Approximation of terrestrial lead isotope evolutionby a two-stage model. Earth and Planetary Science Letters 26 207–221.

Sun, S. & Mcdonough, W. (1989). Chemical and isotopic systematics of oceanic basalts: implications for mantle composition and process. In: *Magnatism in the ocean basins. Geological Society, Special Publication*, Vol.42, pp.313-345, ISSN 0305-8749.

Takehara, L.; Babinski, M.; Toniolo, J.; Chemale Jr., F.; Borba, M. & Guadagnin, F. (2010). Pb-Pb isotope signature of Cu deposits in the Sul-Rio-Grandense Shield, *VII South American Symposium on Isotope Geology*, pp 441-444.

Teixeira, W. (1982). Folhas SH.22-Porto Alegre, SI.22-Lagoa Mirim e SH.21-Uruguaiana: interpretação dos dados radiométricos e evolução geocronológica. Projeto RADAMBRASIL, Florianópolis, Brazil.

Wernick, E.; Hasui, Y. & Brito Neves, B. (1978). As regiões de dobramentos nordeste e sudeste (ext. abst.). *Congresso Brasileiro de Geologia*, Vol. 6, pp. 2493-2506.

Wildner, W.; Lima, E.; Nardi, L. & Sommer, C. (2002). Volcanic cycles and setting in the Neoproterozoic III to Ordovician Camaquã Basin succession in southern Brazil: characteristic of post-collisional magmatism. *Journal of Volcanic and Geothermal Research*, Vol.118, pp.261-283, ISNN 0377-0273.

Wildner W. & Nardi, L. (1999). Características geoquímicas e petrogenéticas do vulcanismo neoproterozóico do sul do Brasil – Platô do Taquarembó – RS. *In*: 1 *Simp. Vulcanismo e Ambientes Associados*, pp.30.

Winchester, J. & Floyd, P. (1977). Geochemical discrimination of different magma series and their differentiation products using immobile elements. *Chemical Geology*, Vol. 20, pp. 325-343, ISSN 0009-2541.

Wood, D. (1980). The application of a Th-Hf-Ta diagram to problems of tectonomagmatic classification and to estabilishing the nature of crustal contamination of basaltic lavas of the British Tertiary volcanic province. *Earth and Planetary Science Letters*, Vol. 50, pp.11-30, ISSN 0084-6597.

Zerfass, H. & Almeida, D. del P. (1997). Mapa geológico da região dos cerros Bugio e Perau, Município de Caçapava do Sul, RS. *Acta Geologica Leopoldensia*, Série Mapas 3, Vol.XX, N°3, pp.1-15, ISNN 0102-1249.

Zerfass, H.; Almeida, D. del P. & Gomes, C. (2000). Faciology of the Acampamento Velho Formation volcanic rocks (Camaquã Basin) in the region of Serra de Santa Bárbara, Cerro do Perau and Cerro do Bugio (Municipality of Caçapava do Sul – RS). *Revista Brasileira de Geociências*, Vol.30, No3, pp.375-379, ISSN 0375-7536.

Zerfass, H.; Almeida, D. del P. & Petry, K. (2001). Mapa Geológico da Serra de Santa Bárbara, Caçapava do Sul (RS): uma contribuição ao conhecimento da Bacia do Camaquã. *Acta Geologica Leopoldensia*, Série Mapas 5, Vol. XXIV, N°5, pp.3-16, ISNN 0102-1249.

Zindler, A. & Hart, S. (1986). Chemical geodynamics. *Earth Planetary Science Letter*, Vol.14, pp.493-571, ISSN 0084-6597.

7

A Combined Petrological-Geochemical Study of the Paleozoic Successions of Iraq

A. I. Al-Juboury
Research Center for Dams and Water Resources, Mosul University, Mosul
Iraq

1. Introduction

Combination of petrographic, mineralogic and geochemical data form the main task of petrologic studies that aim to discuss the provenance history of sedimentary siliciclastic rocks. Provenance analysis serves to reconstruct the pre-depositional history of sediments or sedimentary rocks. This includes the distance and direction of transport, size and setting of the source region, climate and relief in the source area, tectonic setting, and the specific types of source rocks (Pettijohn et al. 1987). Provenance models of sedimentary rocks have, generally taken into account the mineralogical and/or chemical composition of sandstones and shales. Intermingling of detritus from different sources and recycling complicate the determination of sedimentary provenance. Many attempts have been made to refine provenance models using the framework composition and geochemical features (Bhatia and Crook, 1986; Dickinson, 1985; Roser and Korsch, 1988; Zuffa, 1987; Armstrong-Altrin et al., 2004; Umazano et al., 2009 and many others). The chemical composition of the whole rock can provide constraints on provenance because abundance and ratios involving relatively immobile elements are generally not affected by diagenetic processes. Thus chemical data might indicate, in a given sediments, the presence of components which are hard to identify petrographically owing to diagenetic alteration. The geochemical signatures of clastic sediments have been used to find out the provenance characteristics including; the composition of source area (Armstrong-Altrin et al., 2004; Jafarzadeh and Hosseini-Barzi, 2008; Armstrong-Altrin, 2009; Dostal and Keppie, 2009; Umazano et al., 2009; Bakkiaraji etal., 2010), to evaluate weathering processes (Absar et al., 2009; Chakrabarti et al., 2009; Hossain et al., 2010), and to palaeogeographic reconstructions (Ranjan and Banerjee, 2009; Zimmermann and Spalletti, 2009; de Araújo et al., 2010).

The Paleozoic succession of Iraq is exposed in the northernmost part of the country (Fig. 1) and can be traced south and west wards in the subsurface. The Paleozoic succession includes five intracratonic sedimentary cycles, the individual cycles are predominantly siliciclastic, or mixed siliclastic-carbonate units. Sedimentation was mainly controlled by tectonic and eustatic processes which governed the formation of depositional centres, the arrangement of accommodation space within these centres, and the pattern of infilling of the basins (Al-Juboury and Al-Hadidy, 2009). Interbedded sandstones and shales from the Ordovician Khabour Formation and the Devonian-Carboniferous Kaista Formation are selected for this study to evaluate their provenance history.

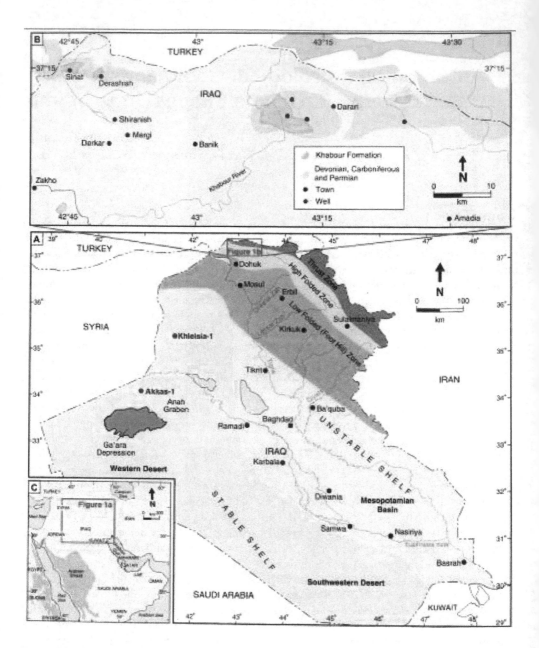

Fig. 1. (a) The structural provinces of Iraq after Buday and Jassim (1987) and the location of the Akkas-1 and Khleisia-1 wells. (b) Paleozoic outcrops in the Ora region including the Khabour and Kaista formations (modified from Al-Omari and Sadiq, 1977). (c) Inset map shows countries neighboring Iraq.

Geochemically derived provenance information from the Paleozoic shales is compared with data from petrographical and geochemical studies of interbedded sandstones and siltstones, in order to assess agreement between the two approaches and to refine knowledge of the provenance for these Paleozoic successions of Iraq.

2. Geologic setting

The stratigraphy of Iraq is strongly affected by the structural position of the country within the main geostructural units of the Middle East region as well as by the structure within Iraq. Iraq lies in the border area between the major Phanerozoic units of the Middle East, i.e., between the Arabian part of the African Platform (Nubio-Arabian) and the Asian branches of the Alpine tectonic belt. The platform part of the Iraqi territory is divided into two basic units, i.e., a stable and an unstable shelf (Figure 1). The stable shelf is characterized by a relatively thin sedimentary cover and the lack of significant folding. The unstable shelf has a thick and folded sedimentary cover and the intensity of the folding increases toward the northeast (Buday 1980). In the Paleozoic, much of the region was covered intermittently by shallow epeiric seas that bordered lowlands made up of portions of the Nubio-Arabian shelf (Al-Sharhan and Nairn 1997). The areal extent of the shelf seas change in response to succeeding transgressions and regressions as the Paleozoic era advanced and their setting varied between tropical and temperate latitudes of the southern hemisphere (Beydoun 1991).

Sedimentary basins of the Paleozoic of Iraq are characterized by the dominance of clastic deposition in the Ordovician and Silurian, with the formation of shallow epeiric seas, which covered large areas of the Arabian Platform. The Arabian Plate represented the northeastern part of the African Plate which extending north and northeastwards over the region now occupied by Iraq, the Arabian Gulf Region, Afghanistan, Pakistan, central, southern, and southeastern Turkey (Numan, 1997). This region represents the northern margin of Gondwana overlooked the southern margins of the Paleo-Tethys Ocean. Epicontinental seas regressed and transgressed over vast areas throughout the Paleozoic, resulting in generally various bed thicknesses and lithotype associations with persistence of facies and absence of unconformities. These characteristics contravene notions (Beydoun, 1991 and Best et al., 1993) that is represented a Gondwana passive margin (Numan, 1997). This region of the Arabian Plate was evolved in AP2 tectonostratigraphic megasequence through intra-cratonic setting (Northern Gondwana land intraplate Paleozoic basin sensu Numan, 1997) with an extension, subsidence and mild uplifting tectonic phase close to Paleo-Tethys passive margin at moderate to high southern latitudes and dominance of clastic sedimentation (Husseini, 1992; McGillivary and Husseini, 1992).

The Paleozoic succession includes five intracratonic sedimentary cycles predominated by siliciclastic, or mixed siliclastic-carbonate units. The Paleozoic cycles commence within the Ordovician with the deposition of the Khabour Formation. This was followed in Silurian times by the Akkas Formation and this is unconformably overlain by the Middle-Late Devonian to Early Carboniferous cycle, represented by the Chalki, Pirispiki, Kaista, Ora and Harur formations. The overlying Permo-Carboniferous cycle is represented by the Ga'ara Formation. The uppermost cycle is late Permian in age and comprises the Chia Zairi Formation. The Paleozoic succession contains a series of muddy units distributed throughout the stratigraphy. The oldest is found in the lower part of the Ordovician Khabour Formation and comprises up to c. 600 m of black fissile shales. Shale units are also present elsewhere within the Ordovician succession, although here they are generally interbedded with sandstones and siltstones. Calcareous shale alternates with sandstone and few dolomites in the Famenian Kaista Formation.

The black shales near the base of the Khabour Formation in western Iraq were also recognized as a maximum flooding surface within the middle part of the Hiswah Formation in Jordan, near the base of the Swab Formation in Syria, and near the base of Saih Nihayda Formation in Oman (Sharland et al. 2001). It is also recognized by Al-Sharhan and Nairn (1997) as a major regional maximum flooding surface separating the Sauk and Tippecanoe sequences sensu Sloss (1963). Lithofacies analysis of the succession in the well Akkas-1 from the western desert of Iraq (Al-Juboury and Al-Hadidy, 2009) revealed that five lithofacies can be recognized. These are; basinal shale facies, transition (shelf to shore-face) facies, tidal storm regressive and transgressive facies, and the near-shore facies.

In surface section of extreme north Iraq, the Khabour Formation consists of alternations of thin-bedded, fine-grained sandstones, quartzites (Cruziana-rich) and silty micaceous shales, olive-green to brown in color. The quartzites are generally cross-bedded, both finely and coarsely, the thicker beds being generally white in color. Bedding planes are usually well-surfaced with smooth films of greenish micaceous shales. Quartzite beds are occasionally truncated by the overlying beds and show fucoids markings, in filled trails and burrows, pitted surfaces and, other bedding-plane structures of unknown origin. Metamorphism is very slight in the thin-bedded shales with quartzites, and almost unnoticeable in the thicker shale beds, (van Bellen et al., 1959). Karim (2006) has noted that the formation in north Iraq was deposited in a spectrum of environments including fluviatile, deltaic, shelf, slope, and deep marine. The depositional environment of the Kaista Formation is interpreted to be a mixed fluvial-marine system. The lower part of the Kaista Formation represents the continuation of clastic influx from the former regressive sequences of the Pirispiki Formation (early Late Devonian), followed by a transgressive phase characterized by a shale facies with glauconite and thin dolostones (Al-Juboury and Al-Hadidy. 2008).

3. Materials and methods

Sandstones and shale samples were selected from the Paleozoic Khabour and Kaista formations from west and North Iraq (Figs. 2 and 3). Totally 50 samples were collected and 24 sandstone (medium to coarse-grained) samples were studied for modal analysis. Between 300-350 grains were counted in each thin section using the Gazzi-Dickinson method to minimize the dependence of rock composition on grain size. Framework parameters (Ingersoll & Suczek, 1979) and detrital modes of sandstones from the studied formations are given in Table 1.

Whole-rock chemical analyses were performed for 28 samples, which include 16 sandstone and 12 shale. Analyses were performed by X-Ray Fluorescence (XRF) and inductively couple plasma-mass spectrometry (ICP-MS) at laboratories of Earth Science Department of Royal Holloway of London University, UK and the results are provided in Tables 2 and 3 respectively. Some X-Ray diffraction (XRD) and scanning electron microscope (SEM) analyses were done at laboratories of Wollongong University (Australia) and Bonn University (Germany).

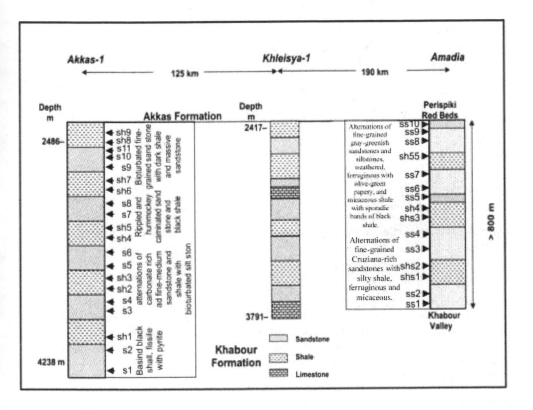

Fig. 2. Generalized lithological succession of the Khabour Formation in Akkas-1 and Khleisya wells of west Iraq and outcrop section at Amadia on north Iraq showing lithological description and location of the analyzed samples.

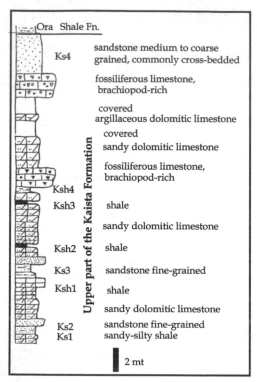

Fig. 3. lithological section of the upper part of the Kaista Formation at Ora region of extreme north Iraq

4. Results

4.1 Sandstone petrography

Quartz is the most dominant constituent of the studied Khabour and Kaista sandstones. Mono-crystalline quartz is the most abundant framework grains. The monocrystalline quartz grains with or without inclusions, the most common inclusions recognized are vacuoles, acicular rutile, spherulitic zircon, muscovite, apatite and iron oxides. Straight to slightly undulatory extinction is frequent type in the quartz studied. According to the genetic and empirical classification of the quartz types (Folk, 1974), the monocrystalline quartz grains are dominantly plutonic and polycrystalline quartz grains are recrystallized and stretched metamorphic types. Sedimentary (Ls), metasedimentary (metamorphic, Lm), and volcanic lithics (Lv), occur in few and varying proportions throughout the sequences of the Khabour and Kaista sandstones (Figs. 4 and 5). Sedimentary lithics (Ls) are the major rock fragments and are dominantly chert. The feldspars are dominated by plagioclase, untwined orthoclase, and twinned microcline (cross-hatching). Mica commonly observed in the studied sandstones in forms of mica laths and biotite. All samples contain accessory minerals, in minor or trace amounts. The dominant heavy minerals identified are zircon, tourmaline, and rutile. Framework composition of the studied Paleozoic sandstones varies from litharenite (sublitharenite, chertarenite) to subarkose and few quartzarenites (Fig. 6). The sandstones are generally cemented by carbonates, secondary silica, ferruginous, and clayey materials.

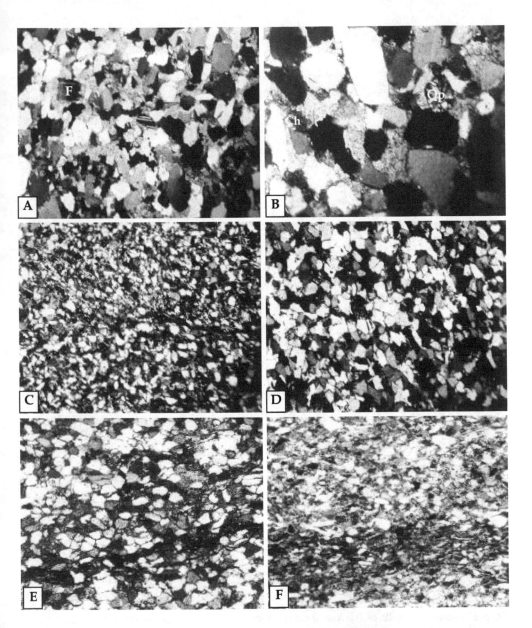

Fig. 4. Photomicrographs of the Khabour sandstones showing (a), monocrystalline quartz and fresh feldspar (F) in carbonate cemented medium grained sandstone. (b), polycrystalline quartz (Qp) and chert (Ch) in medium grained sandstone, note the corroded edges of quartz grains (c), fine-grained sanstones with mica laminations (d), fine-medium grained sandstone, pure quartzarenite with very rare calcite cement patches (e), ferruginous medium grained sandstone (f) fine-grained poorly sorted micaceous sandstone

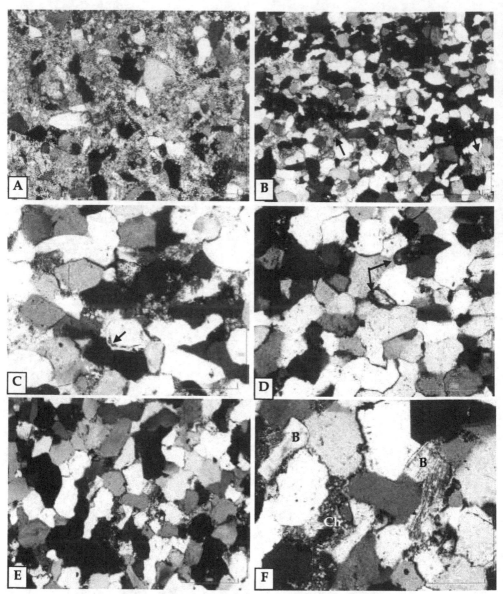

Fig. 5. Photomicrographs of the kaista sandstones showing (A), monocrystalline quartz grains floating in carbonate cement, (B), sandstone with patchy carbonate cement (arrows), (C), iron oxides (sulphides) scattered in quartz rich sandstone, note secondary quartz overgrowth over detrital quartz grain with a chlorite rim between them, (D), highly compacted quartzarenite, note the sutured contacts between grains and two common zircon heavy mineral grains (arrows), (E, and enlarged view in F), compacted sandstone with long-tangential contacts , note chert grains (Ch) and common biotite (B).

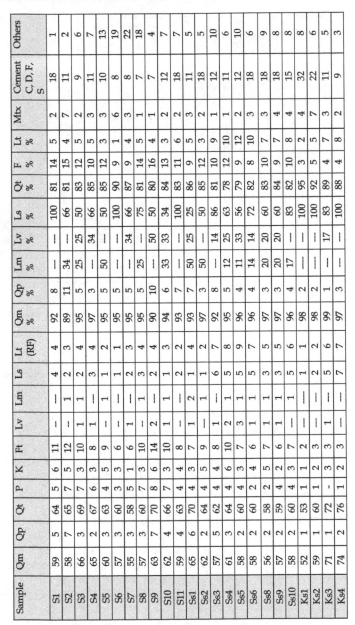

Sample	Qm	Qp	Qt	P	K	Ft	Lv	Lm	Ls	Lt (RF)	Qm %	Qp %	Lm %	Lv %	Ls %	Qt %	F %	Lt %	Mtx	Cement C,D,F,S	Others
S1	59	5	64	5	6	11	—	—	4	4	92	8	—	—	100	81	14	5	2	18	1
S2	58	7	65	7	5	12	—	1	2	3	89	11	34	—	66	81	15	4	7	11	2
S3	66	3	69	7	3	10	1	1	2	4	95	5	25	25	50	83	12	5	2	9	6
S4	65	2	67	6	3	8	1	—	3	4	97	3	—	34	66	85	10	5	3	11	7
S5	60	3	63	4	5	9	—	1	1	2	95	5	50	—	50	85	12	3	3	10	13
S6	57	3	60	3	3	6	—	—	1	1	95	5	—	—	100	90	9	1	6	8	19
S7	55	3	58	5	1	6	—	—	2	3	95	5	—	34	66	87	9	4	3	8	22
S8	57	3	60	7	3	10	1	—	3	4	95	5	25	—	75	81	14	5	1	7	18
S9	63	7	70	8	6	14	—	1	2	4	90	10	—	50	50	80	16	4	1	7	4
S10	62	4	66	7	3	10	2	—	1	3	94	6	33	33	34	84	13	3	2	12	7
S11	59	4	63	4	4	8	1	1	2	2	93	7	—	—	100	83	11	6	2	18	7
Ss1	65	6	70	4	3	7	—	—	1	4	93	7	50	25	25	86	9	5	3	11	5
Ss2	62	2	64	4	5	9	1	2	1	2	97	3	50	—	50	85	12	3	2	18	5
Ss3	57	5	62	4	4	8	2	1	6	7	92	8	—	14	86	81	10	9	1	12	10
Ss4	61	3	64	4	6	10	3	—	5	8	95	5	12	25	63	78	12	10	2	11	6
Ss5	58	2	60	4	3	7	1	1	5	9	96	4	11	33	56	79	9	12	3	12	10
Ss6	58	2	60	2	4	6	1	1	5	7	96	4	14	14	72	82	8	10	3	18	6
Ss8	56	2	58	2	5	7	1	1	3	5	97	3	20	20	60	83	10	7	4	18	9
Ss9	57	2	59	2	2	6	1	1	3	5	97	3	20	20	60	84	9	7	4	18	8
Ss10	58	2	60	4	3	7	1	1	5	6	96	4	17	—	83	82	10	8	4	15	8
Ks1	52	1	53	1	1	2	—	—	1	1	98	2	—	—	100	95	3	2	4	32	8
Ks2	59	1	60	1	2	3	—	—	2	2	98	2	—	—	100	92	5	5	7	22	6
Ks3	71	1	72	—	3	3	—	—	5	6	99	1	—	17	83	89	4	7	3	11	5
Ks4	74	2	76	1	2	3	1	—	7	7	97	3	—	—	100	88	4	8	2	9	3

Table 1. Detrital and authigenic modes of 24 selected samples of Khabour and Kaista sandstones. Qm, monocrystalline quartz, Qp, polycrystalline quartz, Qt, total quartz, P, plagioclase, K, K-feldspar, Ft, total feldspar, Lv, igneous rock fragments, Lm, metamorphic rock fragments, Ls, sedimentary (chert) rock fragments, Lt, total (labile) rock fragments, Mtx, matrix, Cements (C, calcite, D, dolomite, F, ferruginous, S, sericite and illite), others mostly iron oxides, sulphides and heavy minerals.

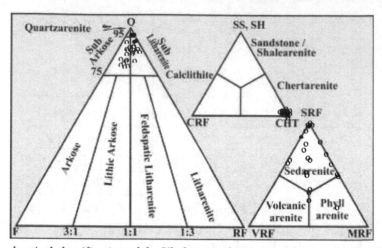

Fig. 6. Minerlaogical classification of the Khabour and Kaista sandstones (Folk, 1974). Q, total quartz; F, Feldspar; RF, rock fragments, SRF, sedimentary rock fragments; VRF, volcanic (igneous rock fragments); MRF, metsedimentary (metamorphic) rock fragments; CHT, chert; CRF, carbonate rock fragments ; SS, sandstone, and SH, shale.
Open circles represents Khabour sandstones and solid circles are Kaista sandstone samples

4.2 Geochemistry
4.2.1 Major elements
Major element distribution reflects the mineralogy of the studied samples. Sandstones are higher in SiO_2 content than shales (Tables 2 and 3 and Fig. 7). Similarly, shales are higher in Al_2O_3, K_2O, Fe_2O_3 and TiO_2 contents than sandstones, which reflect their association in clay-sized phases (Cardenas et al., 1996; Madhavaraju and Lee, 2010). The Al_2O_3 abundances are used as normalization factor to make possible the comparison between different lithologies as it is likely to be immobile during weathering, diagenesis, and metamorphism (Bauluz et al., 2000). In Fig. 7, major oxides are plotted against Al_2O_3. Average UCC(Upper Continental Crust) and PAAS (post Archaean Australian shale) values (Taylor and McLennan, 1985) are also included for comparison. Among other major elements Fe_2O_3, MgO, K_2O, TiO_2 and P_2O_5 are consequently showing strong positive correlations with Al_2O_3, whereas CaO, Na_2O and MnO do not have any trend (Fig. 7). This, strong positive correlations of major oxides with Al_2O_3 indicate that they are associated with micaceous/clay minerals.
The studied samples are normalized to UCC (Taylor and McLennan, 1985) and are given in Fig 8. In comparison with UCC the concentrations of most major elements in sandstones are generally similar, except for Na_2O, with consistently lower average relative concentration value specially for the Kaista sandstones. The depletion of Na_2O (< 1%) in sandstones can be attributed to a relatively smaller amount of Na-rich plagioclase in them, consistent with the petrographic data. K_2O and Na_2O contents and their ratios (K_2O/Na_2O > 1) are also consistent with the petrographic observations, according to which K-feldspar dominates over plagioclase feldspar and common presence of mica as veinlets and patchy distribution in the sandstones of the Khabour Formation (Fig. 4). Some of Kaista sandstones are enriched in CaO and MgO due to the presence of diagenetic calcite and dolomite cements.
In comparison with UCC, the studied shales are low in CaO and Na_2O contents and high in Al_2O_3, K_2O, and TiO_2 contents. Whereas, Kaista shales are enriched in Fe_2O_3 in comparison

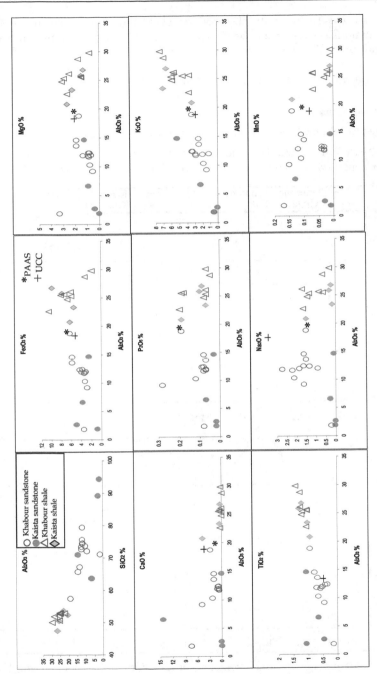

Fig. 7. Major elements versus Al₂O₃ graph showing the distribution of samples from the khabour and Kaista formations. Average data of UCC and PAAS (Taylor and McLennan, 1985) are also plotted for comparison.

with UCC. Al and Ti are easily absorbed on clays and concentrate in the finer, more weathered materials (Das et al., 2006). K_2O enrichment relates to presence of illite as common clay mineral in the studied shales (Fig. 9). On average, the studied shales have lower SiO_2 abundances relative to UCC therefore the observed variations are may be due to quartz dilution effect (Bauluz et al., 2000; Dokuz and Tanyolu, 2006).

4.2.2 Trace elements

4.2.2.1 Large ion lithophile elements (LILE): Rb, Ba, Sr, Th, and U

On average, except Rb all studied sandstones and shales are depleted in Ba, Sr, while they have higher content of Th, and U as compared with UCC (Fig. 8). Th and U show similar

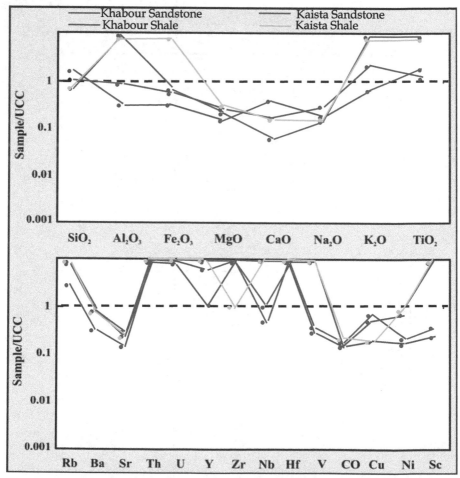

Fig. 8. Spider plot of major and trace elements composition for the Khabour and Kaista sandstones and shales normalized against UCC (Taylor and McLennan, 1985). The trace elements ordered with the large ion-lithophile (LILE) on the left (Rb-U), followed by high field strength elements (HFSE) on the right (Y-Hf) and the transition metals (V-Sc).

geochemical behavior due to their high positive correlation coefficient (r = 0.65; n=16 and r = 0.7; n=12) for sandstones and shales respectively. Except for U and Th, the remaining LILE of the studied Khabour and Kaista sandstones have significant correlations with Al_2O_3. The trace elements such as Sr, Rb, and Ba are correlated positively (r = 0.50, r = 0.60 and r = 0.73, respectively; n=28) against Al_2O_3. These correlations suggest that their distribution is mainly controlled by phyllosilicates. Th weak positive correlation with Al_2O_3 but have strong positive correlations with other elements, such as Ti and Nb (r = 0.72 and r = 0.76, respectively; n=28), implying that it may be controlled by clays and/or other phases (e .g. Ti- and Nb-bearing phases) associated with clay minerals. Rb and Ba are strong positively correlated (r = 0.89; n=16) in sandstones indicating a similar geochemical behavior, and they are also well correlated with K_2O (r = 0.90 and r = 0.89, respectively; n=16). These correlations suggest that their distributions are mainly controlled by illites.

4.2.2.2 High field strength elements (HFSE): Y, Zr, Nb, and Hf

The HFSE elements are enriched in felsic rather than mafic rocks (Bauluz et al., 2000). The concentrations of relatively all HFSE are much higher than UCC (Fig. 8). The well positive correlations for the studied sandstones obtained for TiO_2 with Zr (r = 0.59; n=20), Nb (r = 0.78; n=16), and Hf (r = 0.63; n=16) suggest that their behavior is mainly controlled by the detrital heavy mineral fraction. Zr and Hf behave similar as showed by their high positive correlation coefficient value (r = 0.90; n=16). The Zr/Hf ratio in the analyzed samples ranges from ~ 25-45. This suggests that these elements are controlled by zircons, since these values are nearly identical to those reported by Murali et al. (1983) for zircon crystals. Mean Zr content in shales are lower than the associated sandstones, which indicate that the mineral zircon tends to be preferentially concentrated in coarse-grained sands. These differences between shales and sandstones indicate that sedimentary process such as mineral sorting has played an important role.

4.2.2.3 Transition trace elements (TTE): V, Co, Cu, Ni, and Sc

TTE in the studied sandstones and shales are depleted in comparison with UCC (Fig. 8) except Sc which is more than UCC in shales. The transition trace elements do not behave uniformly. Among TTE, Sc is correlated positively with Al_2O_3 (r = 0.8; n=16) where others are well correlated in sandstones, which indicates that it is mainly concentrated in the phyllosilicates.

4.2.2.4 Rare earth elements (REE)

The ΣREE concentrations of the Khabour and Kaista sandstones are generally lower or nearly same than that of UCC. However, Khabour and Kaista shales are higher than those of UCC. Generally the studied sandstones have less content of REE than shales (ΣREE = 182.2, 281.8 and 126.0, 292.3 for the sandstones and shales of the Khabour and Kaista formation respectively). REE are generally reside in minerals like zircon, monazite, allanite, etc (McLennan, 1989). High REE in Kaista sandstones is due to high zircon content. However, the liner correlation coefficients between ΣREE and Al_2O_3 suggest that clays are also important in hosting the REE (Condie, 1991). If LREE, MREE and HREE are separately considered, all of them show positive correlations with Al_2O_3 (r = 0.48, 0.39 and 0.40, respectively; n=16) and weak positive correlation with Zr.. These positive correlations seem to indicate the variable influence of mineral phases such as phyllosilicates and less effect of zircon in controlling the REE contents.

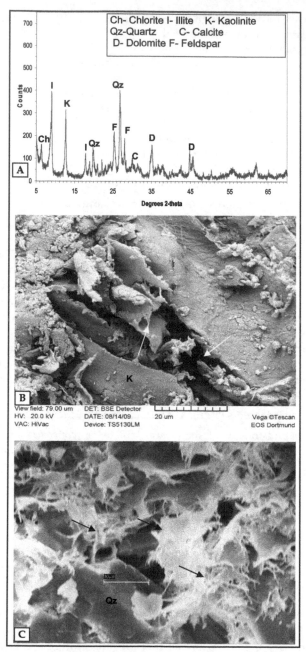

Fig. 9. A- X-Ray diffractogram showing the main clay and non-clay minerals content. B-SEM image illustrating the illite fibers (arrows) and degraded kaolinite hexagonal (K) in the Kaista shale. C- common illite fibers and flakes (arrows) filling pores in Khabour sandstone, Qz is quartz with secondary overgrowth.

Sample	S1	S2	S3	S5	S6	S8	S9	S11	Ss1	Ss3	Ss6	Ss8	Sh1	Sh3	Sh6	Sh8	Shs1	Shs2	Shs3	Shs5
SiO2	71.2	72.8	73.5	75.7	74.6	65.9	67.3	74.5	73.8	72.2	57.5	79.5	52.9	53.1	52.8	52.2	51.9	53.0	50.7	50.1
TiO2	0.14	0.42	0.35	0.55	0.59	0.79	0.69	0.71	0.65	0.45	0.93	0.51	1.07	1.23	1.20	1.27	1.28	1.05	1.03	1.43
Al2O3	1.89	12.41	12.45	11.98	11.73	14.60	13.63	11.89	10.33	9.23	18.81	12.04	22.6	24.7	25.3	26.1	28.7	25.6	25.5	29.8
Fe2O3	3.49	4.31	3.80	3.91	3.43	5.75	5.81	3.27	3.20	2.83	6.20	3.09	10.5	6.72	5.99	6.01	3.23	7.33	8.04	1.89
MnO	0.17	0.03	0.04	0.04	0.03	0.11	0.10	0.03	0.11	0.15	0.14	0.12	0.07	0.03	0.02	0.02	0.01	0.07	0.07	0.01
MgO	3.32	0.86	0.80	1.34	0.83	1.91	1.88	0.90	0.77	0.54	1.67	0.70	2.66	3.12	2.91	2.48	1.75	1.49	1.46	0.82
CaO	7.57	0.95	0.74	1.08	0.66	2.01	1.98	0.76	2.32	4.98	2.86	0.74	0.40	0.12	0.13	0.13	0.05	0.57	0.79	0.36
Na2O	0.23	1.34	1.65	1.86	2.21	1.63	1.51	2.73	2.12	1.63	1.51	0.92	1.67	1.45	1.36	1.71	0.67	0.87	0.65	0.32
K2O	0.52	3.26	3.39	2.92	2.89	2.53	2.43	2.01	1.88	1.48	3.40	1.19	3.80	5.95	6.01	5.70	6.98	3.73	4.59	7.46
P2O5	0.08	0.09	0.08	0.07	0.08	0.08	0.07	0.07	0.12	0.28	0.19	0.07	0.20	0.08	0.07	0.07	0.05	0.18	0.19	0.07
L.O.I.	10.9	2.89	2.71	4.90	2.49	4.51	4.04	2.78	4.48	5.29	6.35	1.12	4.23	3.80	4.03	4.41	4.74	4.98	5.97	6.87
SUM	99.51	99.36	99.51	99.45	99.54	99.82	99.44	99.65	99.78	99.06	99.58	100.0	100.1	100.3	99.82	100.1	99.36	98.87	98.99	99.13
CIA	18.5	69.0	68.3	67.1	67.1	70.3	68.8	68.4	62.0	53.3	70.8	69.5	79.4	76.7	77.1	77.6	78.8	79.9	80.9	78.5
Ni	3.3	18.7	17.9	15.3	16.2	35.0	28.3	15.2	13.5	29.0	39.9	17.7	37.9	38.5	39.3	36.4	40.0	48.9	44.2	49.3
Co	0.2	6.0	5.8	5.0	4.6	14.6	10.3	4.3	7.8	9.2	15.3	8.1	4.8	6.6	15.0	12.6	8.9	5.7	5.6	6.3
Cr	8.8	35.1	34.7	29.4	38.1	70.5	55.2	45.9	29.6	79.9	104.8	12.2	91	100.8	163.1	147.1	147.6	81.7	169.3	144.5
V	13.1	53.3	53.7	41.9	47.4	96.4	76.9	63.0	38.5	66.8	129.1	13.2	85.3	83.8	133.1	181.4	176.9	169.3	106.4	160.5
Sc	1.2	7.1	7.0	6.2	7.2	12.2	9.7	7.2	4.4	11.9	20.0	4.2	16.1	12.3	19.6	17.6	28.5	15.2	23.5	27.6
Cu	5.1	17.2	12.1	21.7	15.1	27.0	19.2	13.7	32.6	30.7	27.1	18.4	39.5	23.5	20.7	29.6	18.3	7.1	11.3	2.8
Zn	223.1	65.1	63.5	134.5	53.3	90.4	86.5	79.6	108.5	99.4	80.7	309.6	73.2	81.4	65.7	51.9	37.7	204.4	57.6	24.0
Pb	12.4	36.8	53.8	35.4	29.1	23.1	20.2	26.5	17.4	12.4	36.9	38.8	8.3	14.3	14.6	10.9	12.5	24.7	53.4	9.7
Sr	70.4	193.3	217.9	203.2	203.4	166.1	175.3	325.6	249.0	70.4	193.4	49.0	177	162	187	224	128.3	120.5	139.6	116.1
Rb	15.7	97.1	100.7	81.1	83.7	99.9	69.5	47.7	134.9	15.7	96.7	12.0	257.1	141.4	176.3	210.1	149.8	153.0	147.9	195.8
Ba	91.2	689.8	700.3	653.6	654.3	448.0	472.6	358.1	765.2	91.2	689.9	122.2	1014	622	657	1316	231	433	831	649
Zr	140.1	179.0	111.4	509.4	415.1	223.4	511.8	488.6	240.6	140.1	178.8	155.5	215.3	176.0	230.3	300.8	318	184	835	282
Hf	4.1	5.1	3.2	12.1	10.0	6.2	13.1	9.2	6.5	13.0	9.3	5.4	5.3	4.2	6.1	9.1	9.4	7.2	19.1	8.1
Nb	2.7	8.9	7.5	10.3	11.3	17.4	15.4	9.6	21.0	2.7	8.9	4.2	26.3	19.1	20.4	27	34.3	21.8	27.2	35.2
Ta	2.5	1.0	1.9	2.1	2.0	2.3	2.6	2.9	2.5	2.5	1.4	2.4	1.0	0.9	1.1	1.7	3.3	3.4	2.9	3.1
Th	2.4	7.4	6.5	12.5	11.3	11.8	12.9	12.1	17.3	2.4	7.4	19.8	23.5	12.9	14.7	22.9	27.3	16.4	26.0	31.3
U	1.1	1.5	1.7	2.7	2.6	2.7	3.3	2.9	3.9	1.1	1.5	1.0	0.9	0.9	0.9	1.6	9.5	6.6	7.8	14.6
Y	21.5	19.7	16.4	21.0	24.4	26.0	30.4	26.1	38.1	21.5	19.7	20.0	30.6	21.8	22.2	29.2	57.8	40.1	38.5	35.7
La	12.0	23.6	26.6	32.0	38.5	27.7	35.9	27.5	39.9	51.9	44.0	44.5	59.2	24.2	49.4	48.6	48.8	61.2	59.1	60.5
Ce	25.4	46.5	52.5	59.5	71.2	54.2	69.8	52.8	87.1	107.8	95.6	87.3	106.6	47.0	81.7	80.1	98.8	116.2	113.4	114.8
Pr	3.0	5.4	6.1	6.7	8.1	6.2	7.9	5.8	9.7	12.1	10.7	9.9	11.8	5.3	10.4	10.3	11.1	13.0	12.3	12.5
Nd	12.3	23.7	25.6	24.0	23.3	30.5	27.1	22.0	48.6	12.3	23.7	42.8	67.4	70.2	71.4	69.6	76.8	52.0	49.7	62.7
Sm	2.1	4.7	4.4	4.9	4.8	3.8	6.2	2.7	7.7	2.1	4.7	10.1	13.2	16.7	15.3	14.2	18.1	13.0	9.1	13.3
Eu	1.0	1.2	1.2	1.2	1.2	1.1	1.1	0.9	1.7	2.1	2.1	1.9	1.7	0.6	2.0	2.0	1.9	2.0	1.7	1.6
Gd	4.8	4.6	4.5	5.2	5.4	4.6	5.4	4.4	6.8	8.6	9.4	10.1	9.2	2.7	9.0	8.9	8.5	8.2	6.9	6.7
Tb	0.9	0.9	0.7	0.9	0.9	0.8	1.0	0.8	1.1	1.5	1.6	1.9	1.7	0.5	1.6	1.6	1.4	1.4	1.2	1.1
Dy	4.1	4.0	3.5	4.0	4.7	4.3	5.1	4.3	4.9	7.6	8.0	8.9	9.2	2.8	8.4	8.3	7.0	7.0	6.4	5.5
Ho	0.8	0.8	0.6	0.8	0.9	0.9	1.0	0.8	0.9	1.6	1.6	1.7	1.9	0.6	1.8	1.8	1.4	1.4	1.4	1.2
Er	1.9	2.4	1.8	2.3	2.6	2.4	3.1	2.5	2.5	4.4	4.3	4.6	5.1	1.8	5.1	4.9	3.7	4.0	4.0	3.4
Tm	0.3	0.3	0.3	0.3	0.4	0.4	0.5	0.4	0.4	0.7	0.6	0.7	0.8	0.3	0.8	0.8	0.6	0.6	0.6	0.5
Yb	1.8	2.0	2.1	2.3	2.9	2.8	3.5	2.7	4.6	4.7	4.6	2.3	5.5	1.9	5.5	5.4	4.0	4.5	4.6	4.0
Lu	0.2	0.3	0.2	0.3	0.4	0.4	0.5	0.4	0.3	0.6	0.6	0.7	0.7	0.3	0.8	0.8	0.6	0.6	0.7	0.9

Table 2. Major and trace elements concentration of selected Khabour sandstone (S and Ss) and shale (Sh) samples. (See Figure 2 for samples location)

Sample	KS1	KS2	KS3	KS4	KSh1	KSh2	KSh3	KSh4
SiO$_2$	94.43	89.26	71.11	63.82	50.58	47.53	53.78	52.5
TiO$_2$	1.07	0.46	1.06	0.65	1.32	1.23	1.02	0.93
Al$_2$O$_3$	1.93	2.71	14.68	6.65	25.84	26.80	23.36	20.81
Fe$_2$O$_3$	0.88	4.99	2.45	3.80	7.83	9.81	5.41	6.20
MnO	0.009	0.029	0.01	0.13	0.01	0.01	0.01	0.14
MgO	0.04	0.40	1.25	0.92	1.51	1.31	2.23	2.67
CaO	0.03	0.08	0.14	14.32	0.41	0.43	0.54	4.86
Na$_2$O	0.04	0.03	0.11	0.28	0.32	0.24	0.18	1.51
K$_2$O	0.49	0.05	5.29	2.27	5.60	6.68	7.08	3.40
P$_2$O$_5$	0.02	0.02	0.03	0.07	0.10	0.09	0.07	0.19
L.O.I.	0.34	1.07	3.37	6.79	6.12	5.29	5.80	6.35
SUM	99.28	99.1	99.5	99.7	99.64	99.42	99.48	99.56
CIA	77.5	91.9	28.6	29.3	80.3	78.5	75.0	68.1
Ni	6.1	22.2	19.8	9.4	48.9	61.9	49.1	39.9
Co	1.1	11.9	2.0	3.5	7.5	8.3	13.8	15.3
Cr	38.0	20.7	86.5	37.2	120.0	129.2	126.5	104.8
V	38.5	15.6	93.1	58.5	124.5	149.8	175.6	129.1
Sc	2.7	1.7	15.6	5.5	25.1	20.9	22.4	20.0
Cu	9.8	7.7	3.4	9.5	3.5	3.7	4.0	27.1
Zn	4.6	22.7	21.5	21.7	71.4	55.7	60.9	80.7
Ga	2.2	2.9	22.4	9.1	28.7	27.7	32.1	13.7
Pb	14.6	8.9	5.2	7.3	12.1	16.6	9.0	36.9
Sr	23.6	28.5	40.5	147.9	108.8	161.3	69.1	193.4
Rb	10.5	2.5	224.3	85.6	184.4	189.5	271.0	96.7
Ba	29.0	31	895	295	542	542	467	495
Zr	1115	343	445	668	306	226	128	178.8
Hf	31	8.0	12	15	8	6	5	9
Nb	21.5	8.9	28.9	14.5	30.0	31.5	27.1	8.9
Ta	2.8	2.0	3.5	1.2	2.4	3.3	2.7	1.4
Th	22.9	9.4	22.5	12.8	27.1	24.2	21.8	7.4
U	4.1	2.1	3.0	1.8	6.0	4.8	5.3	1.5
Y	30.3	15.4	21.5	33.1	38.2	51.8	22.4	19.7
L.a	16.2	21.1	34.1	34.5	31.7	92.1	44.9	100.1
Ce	31.0	42.7	64.2	63.8	55.5	170.1	76.4	193.8
Pr	3.7	5.5	6.7	7.6	6.1	18.1	8.2	20.4
Nd	14.3	23.7	24.3	31.2	22.6	75.0	29.6	84.5
Sm	2.1	5.1	3.4	6.2	3.8	14.2	4.3	15
Eu	3.0	1.0	0.6	1.0	0.8	2.9	0.8	2.7
Gd	0.5	4.5	2.9	5.3	3.5	10.8	3.8	10.9
Tb	2.9	0.8	0.6	1.0	0.7	1.8	0.7	1.6
Dy	0.5	3.8	3.2	4.9	4.0	8.8	3.8	6.9
Ho	2.7	0.7	0.7	1.0	0.9	1.7	0.8	1.3
Er	1.6	2.0	2.4	2.8	2.7	4.8	2.5	3.7
Tm	0.2	0.3	0.4	0.4	0.4	0.7	0.4	0.6
Yb	1.6	1.9	3.0	3.1	3.2	5.2	2.9	4.2
Lu	0.2	0.3	0.5	0.4	0.5	0.8	0.5	0.5

Table 3. Major and trace elements concentration of selected sandstone (KS) and shale (KSh) samples of the Kaista Formation (See Figure 3 for samples location)

5. Provenance information from sandstones

5.1 Source rocks

As pointed out above, sandstones petrographic investigation revealed that they are variable (Table 1) with detrital quartz being the most abundant and constant component. The average quartz content of the Khabour sandstones is 63% and 65% for the Kaista sandstones. The feldspar content range from 6% to 12 %, and from 2% to 3% in the Khabour and Kaista sandstones respectively. Rock fragments range from 1 % to 9% in the Khabour sandstones and from 1% to 7% in the Kaista sandstones with sedimentary rock fragments dominated by chert being the dominant and small and occasional content of igneous and metamorphic fragments.

The qualitative petrography study provides important information on the nature of the source area. Mono-crystalline quartz is the most abundant framework grains. Whereas, few polycrystalline quartz grains of (> 3 grains) per each polycrystalline grain were identified. Most of monocrystalline quartz grains are of straight to slightly undulatory extinction, with or without inclusions; where present, the most common inclusions are vacuoles, acicular rutile, spherulitic zircon, muscovite, apatite and iron oxides. Quartz types, inclusions and undulosity indicate a derivation from a dominantly plutonic (granitic) provenance with subordinate input from low rank metamorphic rocks. (Basu et al., 1975).

to discriminate provenance fields for the studied rocks, a TiO_2 vs. Ni bivariate plot (Fig. 10; Floyd et al., 1989) is used. The majority of samples plot in the acidic field, even though few samples plot outside the field assigned for felsic source.

On a the La/Th vs. Hf bivariate (Fig. 11; Floyd and Leveridge, 1987) suggests the felsic source rocks although there are some differences in source rocks between shales and sandstones. Furthermore, La/Sc versus Th/Co bivariate diagram (Fig. 12; Cullers, 2002), shows that nearly all the studied samples plot near to the silicic rock provenance composition. In addition, the REE patterns and the size of the Eu anomaly have been also used to infer sources of sedimentary rocks (Taylor and McLennan, 1985). Since basic igneous rocks contain low LREE/HREE ratios and little or no Eu anomalies, whereas silicic igneous rocks usually contain higher LREE/HREE ratios and negative Eu anomaly (Cullers, 1994; Cullers et al., 1987). The average chondrite normalized REE patterns of the studied rocks are shown in Figure 13.

For comparison average REE patterns of Continental Crust , Continental Arc, Mid-Oceanic Ridge, and Oceanic Island Basalt are also included in this Figure 13. The chondrite normalized REE patterns for the Khabour and Kaista sandstones and shales are comparable to Continental Crust and Continental Arc. The REE patterns suggest that the samples were mainly derived from an old upper continental crust composed chiefly of felsic components. Similarly, in the Eu/Eu* and Th/Sc plot (Fig. 14; Cullers and Podkovyrov, 2002) the samples plot between the average values of granite and granodiorite source rocks with rare mafic provenance.

The post-Archean pelites have low concentrations of mafic elements, particularly Ni and Cr, when compared to Archean pelites (McLennan et al., 1993). The reason for the high concentrations of Ni and Cr in the Archean pelites is due to the deficiency of ultra-mafic rocks in the post Archaean Period (Taylor and McLennan, 1985). The studied sandstones plot in the post Archaean field (Fig. 15) and suggest that the felsic component was dominant in the source area of the Khabour and Kaista formations. The (Gd/Yb)CN ratio also document the nature of source rocks and the composition of the continental crust (Nagarajan et al., 2007; Taylor and McLennan, 1985). On a Eu/Eu* vs. (Gd/Yb)CN diagram (Fig. 16), the studied shales and most of the sandstones plot in the post Archean field and near to PAAS value, which suggest that the post Archean felsic rocks could be the source rocks for the Khabour and Kaista formations. Archean sources could be compared with those sources recorded for Paleozoic clastics in southern Turkey (Kröner and Sengör, 1990) and Iran (Etemad Saeed etal., 2011).

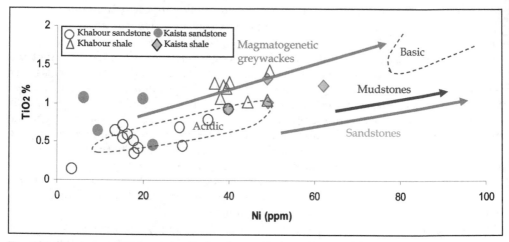

Fig. 10. TiO₂ versus Ni bivariate plot for the studied sandstones (Floyd et al., 1989). Majority of samples plot near the acidic sources.

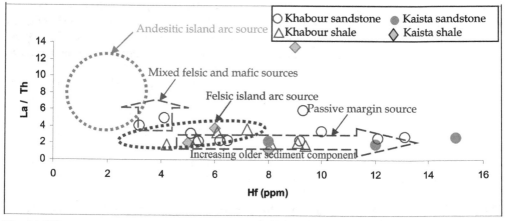

Fig. 11. Hf versus La/Th diagram (Floyd and Leveridge, 1987).

McLennan et al. (1990) recognized four distinctive provenance components on the basis of geochemistry: old upper continental crust, young undifferentiated arc, young differentiated (Intracrustal) arc and Mid-Ocean ridge basalt (MORB). This study reveals that the studied sandstones and shales were derived from an old and well-differentiated upper continental crust provenance, which is characterized by high abundances of large ion lithophile (LILE) elements, high Th/Sc, La/Sm, Th/U ratios and negative Eu anomaly (McLennan et al., 1990). It seems that the felsic source for the Khabour and Kaista formations are similar to the

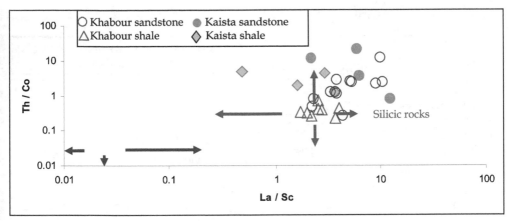

Fig. 12. Th/Co versus La/Sc plot (Cullers, 2002). The studied sandstones and shales plot near the silicic source.

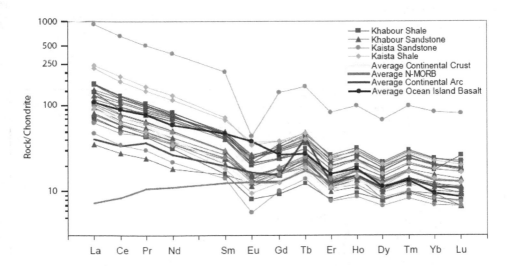

Fig. 13. Chondrite normalized rare earth element plots for the studied sandstones and shales. Average Continental Crust , Continental Arc, Mid-Oceanic Ridge, and Oceanic Island Basalt are also included. Data sources: Average Upper continental crust (Taylor and McLennan, 1995), N-MORB (average Sun and McDonough 1989) Continental arc (average from Georoc database query basaltic andesite convergent margin, ICPMS, REE only), Ocean Island basalt (Sun and McDonough 1989)

acidic and basic igneous basement rocks of Iraq. The crystalline basement rocks of Iraq is interpreted from seismic and geophysical data to range in depth from about 6–10 km and is composed mostly of granitic, basic and ultra basic igneous and metamorphic rocks (Buday, 1980; Al-Hadidy, 2007).

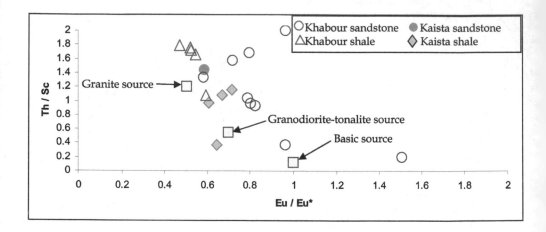

Fig. 14. Eu/Eu*-Th/Sc bivariate plot for the samples from the Khabour and Kaista formations (Cullers and Podkovyrov, 2002).

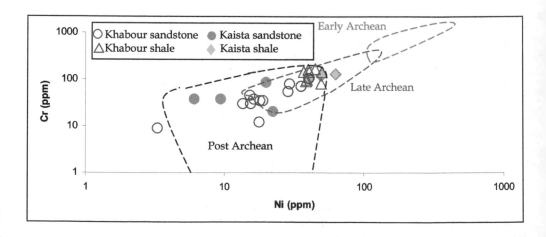

Fig. 15. Ni-Cr bivariate plot for the samples from the Khabour and Kaista formations (McLennan et al., 1993).

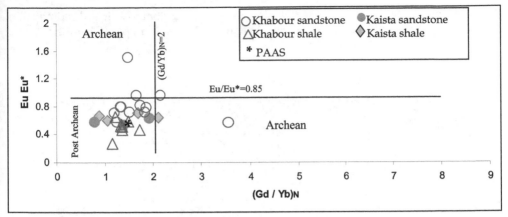

Fig. 16. Plot of Eu/Eu* versus (Gd/Yb)CN for the samples of the studied formations. Fields are after McLennan and Taylor (1991).

5.2 Implications for tectonic setting

Petrographic data from various framework constituents (Quartz, Feldspar, and Rock Fragments) were plotted on various ternary and bivariate diagrams to show their positions on various schemes in order to discriminate their tectonic settings and show their paleoclimatic and weathering conditions. On the Qt-F-L and Qm-F-Lt diagrams (Figure 17A) of Dickinson and Suczek, (1979), the Khabour sandstones plot in the recycled orogen and continental block provenances with stable craton sources and with uplifting in the basement complexes. Whereas, Kaista sandstones were plotted in the recycled Orogen Provenance. Similarly, in the Lm-Lv-Ls and Qp-Lvm-Lsm ternary diagrams of Ingersoll and Suczek (1979) (Figure 17B) the studied sandstones plot mostly in mixed arc and subduction continental margin and in rifted continental margins and partly in sutured belt provenances.

Within recycled orogens, sediment sources are dominantly sedimentary with subordinate volcanic rocks derived from tectonic settings where stratified rocks are deformed, uplifted and eroded (Dickinson, 1985; Dickinson and Suczek, 1979). As pointed out by Dickinson et al. (1983), sandstones plotting in craton interior field are mature sandstones derived from relatively low-lying granitoid and gneissic sources, supplemented by recycled sands from associated platform or passive margin basins. The detrital modal compositions of both Khabour and Kaista sandstones are plotted in the Q-F-L diagram (Fig. 18; Yerino and Maynard, 1984), which indicates that these sandstones are related to trailing-edge margin. Bhatia (1983) and Roser and Korsch (1986) proposed tectonic setting discrimination fields for sedimentary rocks to identify the tectonic setting of unknown basins. These tectonic setting discrimination diagrams are still extensively used by many researchers to infer the tectonic setting of ancient basins (Drobe et al., 2009; Gabo et al., 2009; Maslov et al., 2010; Wani and Mondal, 2010 Bakkiaraji et al., 2010; Bhushan and Sahoo, 2010; de Araújo et al., 2010). However, the functioning of major elements tectonic setting discrimination diagrams proposed by Bhatia (1983) and Roser and Korsch (1986) have been evaluated in many studies. Armstrong-Altrin and Verma (2005) observed that the tectonic setting discrimination diagram proposed by Roser and Korsch (1986) works better than Bhatia's

(1983) diagram. In this study, K₂O/Na₂O versus SiO₂ tectonic setting discrimination diagram (Fig. 19) shows that most of the Khabour and Kaista samples fall in the Active continental and passive margin fields.

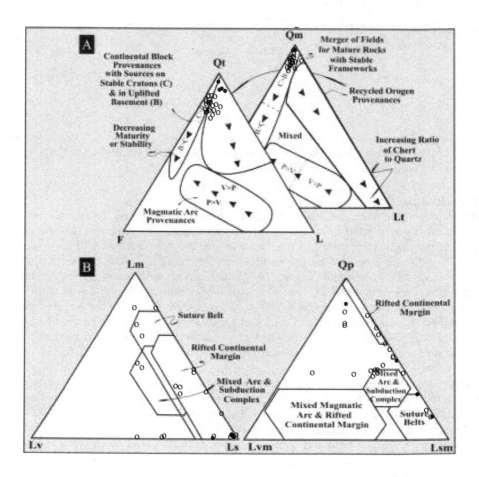

Fig. 17. Provenance diagrams for the studied sandstones (A) Qt-F-L and Qm-F-Lt plots.Tectonic setting fields after Dickinson and Suczek (1979), and (B) Lm-Lv-Ls and Qp-Lvm-Lsm after Ingersoll and Suczek (1979). Data and definitions are given in Table 1.

As discussed above, the Khabour and Kaista sandstones posses similar characteristics of a passive margin setting as described by McLennan et al. (1993). Passive margin sediments are largely quartz-rich, derived from plate interiors or stable continental margins. Bhatia (1983) opined that the sedimentary rocks deposited on passive margins are characterized by enrichment of LREE over HREE with pronounced negative Eu anomaly on chondrite-normalized patterns.

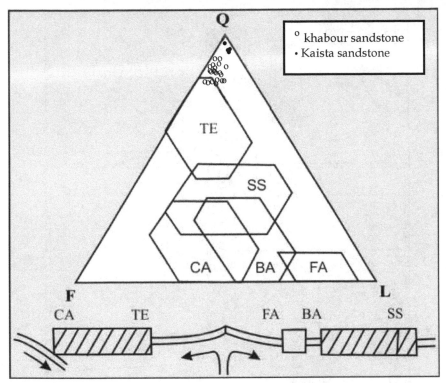

Fig. 18. Q-F-L tectonic provenance diagram for the Khabour and Kaista sandstones, after Yerino and Maynard (1984). The studied sandstones plot near the TE field. TE: trailing edge (also called passive margin); SS: strike-slip; CA: continental-margin arc; BA: back arc to island arc; FA: fore arc to island arc.

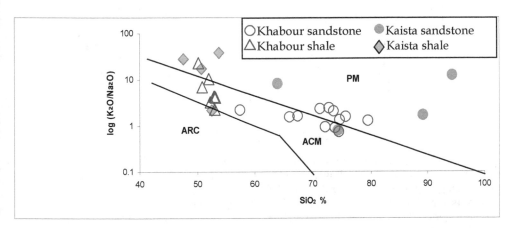

Fig. 19. Tectonic-setting discrimination diagram after Roser and Korsch (1986). PM = passive margin; ACM = Active continental margin; ARC = Island arc.

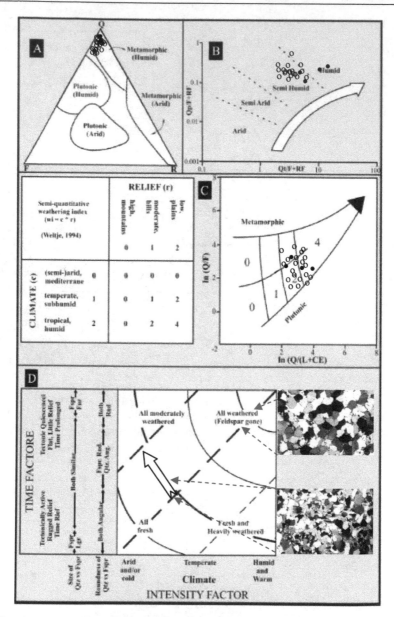

Fig. 20. Illustrating the effect of climate on the composition of the Khabour sandstones using, A- Suttner et al., (1981) diagram. Q:Quartz; F: Feldspar, R: Rock fragments. B- Bivariate log/log plot (Suttner and Dutta, 1986). Qt: total Quartz, F: Feldspar, RF: Rock fragments, Qp: Polycrystalline quartz. C- Weathering diagram and semi-quantitative weathering index after Weltje (1994). CE: Carbonate clasts. D- Evaluate of paleoclimate condition based on relation between quartz and feldspar grains and degree of weathering of feldspar grains (Folk, 1974).

5.3 Weathering, relief, and climate

In the Q-F-R ternary diagram (Suttner et al., 1981), Khabour and Kaista sandstones plot in the field of the metamorphic source area with humid climate (Fig. 20A). In addition, in the bivariate diagram of Suttner and Dutta (1986) the studied sandstones reveal the differences in climate condition from semi-arid to humid (Fig. 20B). Similarly, in the Grantham and Velbel (1988) weathering index wi = c* r and Weltje (1994) diagrams (Fig. 20C), the studied sandstones plot into the field of wi = 2 and 4 indicating moderate to high degree of weathering in low plains relief and from semi-arid to semi-humid climate conditions and mainly between metamorphic and plutonic compositions. Furthermore, in the Folk (1974) weathering intensity diagram (Fig. 20D), some of the Khabour and Kaista sandstones plot in the mixed moderately weathered field and fresh feldspars plot in the temperate to arid climate field, whereas quartzite sandstones of both formations plot in the humid climate field. The intensity and duration of weathering in clastic sediments can be evaluated by examining the relationships among alkali and alkaline rare earth elements (Nesbitt and Young, 1996; Nesbitt et al., 1997). Various investigators have utilized the so-called "Chemical Index of Alteration" (CIA) of Nesbitt and Young (1982) to evaluate the intensity and the degree of chemical weathering: CIA = $[Al_2O_3/(Al_2O_3 + CaO + Na_2O + K_2O)]$ * 100, where the oxides are expressed as molar proportions and CaO represents the Ca in silicate fractions only. The high CIA values in shales (mean 79 and 76, for the Khabour and Kaista formations respectively) and most of the studied sandstones (see Tables 2 and 3) indicate a moderate to intense weathering of first cycle sediment, or alternatively, recycling could have produced these rocks.

6. Conclusions

The Ordovician Khabour Formation in subsurface sections of west Iraq and in surface section of extreme north Iraq consists of sandstones and shales. Whereas, sandstone units of Devonian-Carboniferous Kaista Formation intercalate with limestone and shales The provenance of these formations has been assessed using integrated petrographical and geochemical data of the interbedded sandstones and shales to arrive at an internally consistent interpretation. The Khabour sandstones are subarkose and sublitharenite with few quartzarenite and derived largely from recycled orogen and continental block provenances while Kaista sandstones are mostly quartzarenite from recycled orogen. Both studied sandstones are predominantly derived from a felsic and rare mafic sources with a component from pre-existing sedimentary and volcanic rocks. Compositional differences and increase in the degree of weathering from sandstones to shales indicate climatic variations (semi-arid to humid) in the source area. In general, the acidic (felsic sources) and rare mafic sources with a prevailing continental margin tectonic setting for the Khabour sandstones, in accordance with higher values of Thorium/Scandium (Th/Sc) and Thorium/Uranium (Th/U) values seem that the felsic and mafic sources for the Khabour sandstone are likely consisted of basement rocks of Iraq. The Kaista sandstones were recycled from older sedimentary succession and were deposited in a fluvio-marine depositional system with dominating moderate to high degree of weathering in low plains regions and from semi-arid to semi-humid climate conditions.

7. References

Absar, N., Raza, M., Roy, M., Naqvi, S.M. & Roy, A.K. (2009). Composition and weathering conditions of Paleoproterozoic upper crust of Bundelkhand craton, Central India: Records from geochemistry of clastic sediments of 1.9 Ga Gwalior Group. *Precambrian Research*, Vol.168, No. 3-4, 313-329.

Al-Hadidy, A.H. (2007). Paleozoic stratigraphic lexicon and hydrocarbon habitat of Iraq. *GeoArabia*, Vol. 12, No. 1, 63-130.

Al-Juboury, A.I. & Al-Hadidy, A.H. (2008). Facies and depositional environment of the Devonian-Carboniferous succession of Iraq .Geological Journal, Vol. 43, No. 2-3, 383-396.

Al-Juboury, A.I. & Al-Hadidy, A.H. (2009). Petrology and depositional evolution of the Paleozoic rocks of Iraq, Marine & Petroleum Geology, Vol. 26, No.2, 208-231.

Al-Omari, F.S. & Sadiq, A. (1977). Geology of Northern Iraq, Dar Al-Kutib press, Mosul University, Iraq, 198pp.

Al-Sharhan, A.S. & Nairn, A.E.M. (1997). Sedimentary Basins and Petroleum Geology of the Middle East. Elsevier, Amsterdam, 843pp.

Armstrong-Altrin, J.S. (2009). Provenance of sands from Cazones, Acapulco, and Bahía Kino beaches, México. *Revista Mexicana de Ciencias Geológicas*, Vol. 26, (3), 764-782

Armstrong-Altrin, J.S. & Verma, S.P. (2005). Critical evaluation of six tectonic setting discrimination diagrams using geochemical data of Neogene sediments from known tectonic settings. *Sedimentary Geology*, Vol. 177, 115-129.

Armstrong-Altrin, J.S., Lee, Y.I., Verma, S.P., Ramasamy, S. (2004). Geochemistry of sandstones from the upper Miocene Kudankulam Formation, Southern India: Implications for provenance, weathering, and tectonic setting. *Journal of Sedimentary Research*, Vol.74, No.2, 285-297.

Bakkiaraji, D., Nagendrai, R., Nagarajan, R. & Armstrong-Altrin, J.S. (2010). Geochemistry of sandstones from the Upper Cretaceous Sillakkudi Formation, Cauvery Basin, Southern India: Implication for Provenance. Journal of the Geological Society of India, Vol.76, 453-467.

Basu, A., Young, S., Suttner, L., James, W. & Mack, G.H. (1975). Re-evaluation of the use of undulatory extinction and crystallinity in detrital quartz for provenance interpretation. *Journal of Sedimentary Petrology*, Vol.45, 873-882.

Bauluz, B., Mayayo, M.J., Femandez-Nieto, C. & Gonzalez-Lopez, J.M. (2000). Geochemistry of Precambrian Paleozoic siliciclastic rocks from the Iberian Range (NE Spain): implication for source area weathering, sorting, provenance and tectonic setting. *Chemical Geology*, Vol.168, 135- 150.

Best, J.A., Barazangi, M., Al-Saad, D., Sawaf, T. & Gebran, A.. (1993). Continental margin evolution of the northern Arabian platform in Syria. *American Association of Petroleum Geologists Bulletin*, Vol.77 , No.2, 173-193.

Beydoun, Z.R. (1991). Arabian plate hydrocarbon, geology and potential: A plate tectonic approach, *AAPG Studies in Geology*, 33, 77pp.

Bhatia, M.R. (1983). Plate tectonics and geochemical composition of sandstones. *Geology*, Vol. 91, 611-627.

Bhatia, M.R. & Crook, K.A. W. (1986). Trace elements characteristics of greywackes and tectonic setting discrimination of sedimentary basins. *Contribution to Mineralogy and Petrology*, Vol.92, 181-193.

Bhushan, S.K. & Sahoo, P. (2010). Geochemistry of clastic sediments from Sargur supracrustals and Bababudan Group, Karnataka: implications on Archaean Proterozoic Boundary. *Journal Geological Society of India*, Vol.75, 829-840.

Buday, T. (1980). The Regional of Iraq. Vol.1, Stratigraphy & Palaeogeography. State Organization for Minerals, Baghdad, Iraq, 445pp.

Buday, T. & Jassim, S.Z. (1987). The Regional Geology of Iraq, vol. 2, Tectonism, Magmatism and Metamorphism. Publication of the Geological Survey of Iraq. Vol.2, 352p.

Cardenas, A., Girty, G.H., Hanson, A.D. & Lahren, M.M. (1996). Assessing differences in composition between low metamorphic grade mudstones and high-grade schists using log ratio techniques. *Journal of Geology*, Vol.104, 279-293.

Chakrabarti, G., Shome, D., Bauluz, B. & Sinha, S. (2009). Provenance and weathering history of Mesoproterozoic clastic sedimentary rocks from the Basal Gulcheru Formation, Cuddapah Basin, India. *Journal Geological Society of India*, Vol.74, 119-130.

Condie, K.C. (1991). Another look at rare earth elements in shales. *Geochimica et Cosmochimica Acta*, Vol.55, 2527-2531.

Cullers, R.L. (1994). The controls on the major and trace element variation of shales, siltstones and sandstones of Pennsylvanian-Permian age from uplifted continental blocks in Colorado to platform sediment in Kansas, USA. *Geochimica et Cosmochimica Acta*, Vol. 58, No.22, 4955-4972.

Cullers., R.L. (2002). Implications of elemental concentrations for provenance, redox conditions, and metamorphic studies of shales and limestones near Pueblo, CO, USA. *Chemical Geology*, Vol. 191, No.4, 305-327.

Cullers, R.L. & Podkovyrov, V.N. (2002). The source and origin of terrigenous sedimentary rocks in the Mesoproterozoic Ui group, southeastern Russia. *Precambrian Research* Vol.117, 157-183.

Cullers, R.L., Barret, T., Carlson, R. & Robinson, B. (1987). Rare earth element and mineralogical changes in Holocene soil and stream sediment: a case study in the Wet Mountains, Colorado, USA. *Chemical Geology*, Vol. 63, 275-295.

Das, B.K., AL-Mikhlafi, A.S. & Kaur, P. (2006). Geochemistry of Mansar Lake sediments, Jammu, India: Implication for source-area weathering, provenance, and tectonic setting. *Journal of Asian Earth Sciences*, Vol.26, 649-668.

De Araújo, C.E.G., Pinéo, T.R.G., Caby, R., Costa, F.G., Cavalcante, J.C., Vasconcelos, A.M. & Rodrigues, J.B. (2010). Provenance of the Novo Oriente Group, southwestern Ceará Central Domain, Borborema Province (NE-Brazil): A dismembered segment of a magma-poor passive margin or a restricted rift-related basin?. *Gondwana Research*, Vol.18, 497-510.

Dickinson, W.R. (1985). Interpreting detrital modes of greywacke and arkose. *Journal Sedimentary Petrology*, Vol.40, 695-707.

Dickinson, W.R. & Suczek, C.A. (1979). Plate tectonics and sandstone compositions. *American Association of Petroleum Geologist Bulletin*, Vol. 63, 2164-2182.

Dickinson, W.R., Beard, L. S., Brakenridge, G.R., Evjavec, J.L., Ferguson, R.C., Inman, K.F., Knepp, R.A., Lindberg, F.A. & Ryberg, P.T. (1983). Provenance of North American Phanerozoic sandstones in relation to tectonic setting. *Geological Society of America Bulletin*, Vol.94, 222-235.

Dokuz, A. & Tanyolu, E. (2006). Geochemical constraints on the provenance, mineral sorting and subaerial weathering of lower Jurassic and upper Cretaceous clastic rocks of the eastern Pontides, Yusufeli (Artvin), NE Turkey. *Turkish Journal of Earth Sciences,* Vol.15, 181-209.

Dostal, J. & Keppie, J.D. (2009). Geochemistry of low-grade clastic rocks in the Acatl n Complex of southern Mexico: Evidence for local provenance in felsic-intermediate igneous rocks. *Sedimentary Geology,* Vol.222, 241-253.

Drobe, M., Lpez de Luchi, M.G., Steenken, A., Frei, R., Naumann, R., Siegesmund, S. & Wemmera, K. (2009). Provenance of the late Proterozoic to early Cambrian metaclastic sediments of the Sierra de San Luis (Eastern Sierras Pampeanas) and Cordillera Oriental, Argentina. *Journal of South American Earth Sciences ,* Vol.28, 239-262.

Etemad-Saeed, N., Hosseini-Barzi, M. & Armstrong-Altrin, J.S. (2011). Petrography and geochemistry of clastic sedimentary rocks as evidences for provenance of Lower Cambrian Lalun Formation, Posht-e-badam block, Central Iran, *Journal of African Earth Sciences,* Vol.61, No.2, 142-159.

Floyd, P.A. & Leveridge, B.E. (1987). Tectonic environment of the Devonian Gramscatho basin, south Cornwall: framework mode and geochemical evidence from turbiditic sandstones. *Journal of the Geological Society London ,* Vol.144, 531-542.

Floyd, P.A., Winchester, J.A. & Park, R.G. (1989). Geochemistry and tectonic setting of Lewisianclastic metasediments from the Early Proterozoic Loch Maree Group of Gairloch, N.W. Scotland. *Precambrian Research ,* Vol.45, No. 1-3, 203-214.

Folk, R.L. (1974). Petrology of Sedimentary Rocks, Hemphill Publication Company, Texas, 170pp.

Gabo, J.A.S., Dimalanta, C.B., Asio, M.G.S., Queao, K.L., Yumul Jr, G.P. & Imai, A. (2009). Geology and geochemistry of the clastic sequence from northwestern Panay (Philippines): Implications for provenance and geotectonic setting. *Tectonophysics,* Vol.479, 111-119.

Grantham, J.H. & Velbel, M.A. (1988). The influence of climate and topography on rock fragments abundance in modern fluvial sands of the southern Blue Ridge Mountains, North Carolina, *Journal of Sedimentary Petrology,* Vol.58, 219-227.

Hossain, H.M.Z., Roser, B.P. & Kimura, J.I. (2010). Petrography and whole-rock geochemistry of the Tertiary Sylhet succession, northeastern Bengal Basin, Bangladesh: provenance and source area weathering. *Sedimentary Geology,* Vol.228,171-183.

Husseini, M. I. (1992). Upper Paleozoic tectono-sedimentary evolution of the Arabian and adjoining plates. *Journal of the Geological Society,* London, Vol. 149, 419-429.

Ingersoll R. V. & Suczek C.A. (1979). Petrology and Provenance Neogene sand from Nicobar and Bengal fans. DSDP site 211and 218. *Journal of Sedimentary Petrology.* Vol.49. 1217-1228.

Jafarzadeh, M. & Hosseini-Barzi, M. (2008). Petrography and geochemistry of Ahwaz sandstone member of Asmari Formation, Zagros, Iran: implications on provenance and tectonic setting. *Revista Mexicana de Ciencias Geolgicas,* Vol.25, No.2, 247-260.

Karim, K.H. (2006). Comparison study between the Khabour and Tanjero Formations from North Iraq. Iraqi Journal of Earth Science, Mosul University,Vol. 6, No. 2, 1-12.

Kröner, A. & Sengör, A.M.C. (1990). Archean and Proterozoic ancestry in late Precambrian to early Paleozoic crustal elements of southern Turkey as revealed by single-zircon dating. *Geology*, Vol. 18 (12), 1186-1190.

Madhavaraju, J. & Lee, Y.I. (2010). Influence of Deccan volcanism in the sedimentary rocks of Late Maastrichtian–Danian age of Cauvery basin Southeastern India: constraints from geochemistry. *Current Science* , Vol.98, No.4, 528-537.

Maslov, A.V., Gareev, E.Z. & Podkovyrov, V.N. (2010). Upper Riphean and Vendian sandstones of the Bashkirian anticlinorium. *Lithology and Mineral Resources*, Vol. 45, No. 3, 285-301.

McGillivray, J.G. & Husseini, M.I. (1992). The Paleozoic petroleum geology of central Arabia. *American Association of Petroleum Geologists Bulletin*, Vol. 76, 1473-1490.

McLennan, S.M. (1989). Rare earth elements in sedimentary rocks: influence of provenance and sedimentary processes, in: Lipin, B.R., McKay G.A. (Eds.), Geochemistry and mineralogy of rare earth elements. *Reviews in Mineralogy*, Vol.21,169-200.

McLennan, S.M., & Taylor, S.R. (1991). Sedimentary rocks and crustal evolution, Tectonic setting and secular trend, *Journal of Geology*, Vol. 99, 1-21

McLennan, S.M., Taylor, S.R., McCulloch, M.T. & Maynard, J.B. (1990). Geochemical and Nd-Sr isotopic composition of deep-sea turbidites: Crustal evolution and plate tectonic associations. *Geochimica et Cosmochimica Acta*, Vol.54, 2015-2050.

McLennan, S.M., Hemming, S., McDaniel, D.K. & Hanson, G.N. (1993). Geochemical approaches to sedimentation, provenance and tectonics, in: Johnsson, M.J. and Basu, A. (Eds.), Processes Controlling the Composition of Clastic Sediments. *Geological Society of American Special Paper*, 21–40.

Murali, A.V., Parthasarathy, R., Mahadevan, T.M. & Sankar Das M. (1983). Trace element characteristics, REE patterns and partition coefficients of zircons from different geological environment- a case study on Indian zircons. *Geochimica et Cosmochimica Acta*, Vol.47, 2047-2052.

Nagarajan, R., Madhavaraju, J., Nagendra, R., Armstrong-Altrin, J.S. & Moutte, J. (2007). Geochemisrty of Neoproterozoic shales of the Rabanpalli Formation Bhima Basin, Northern Karnataka, southern India: implications for provenance and paleoredox conditions. *Revista Mexicana de Ciencias Geologicas* , Vol.24, 20-30.

Nesbitt, H.W. & Young, G.M. (1982). Early Proterozoic climates and plate motions inferred from major element chemistry of lutites. *Nature* , Vol.299, 715-717.

Nesbitt, H.W. & Young, G.M. (1996). Petrogenesis of sediments in the absence of chemical weathering: effects of abrasion and sorting on bulk composition and mineralogy. *Sedimentology*, Vol.43, 341-358.

Nesbitt, H.W., Fedo, C.M. & Young, G.M. (1997). Quartz and feldspar stability, steady and non steady state weathering and petrogenesis of siliciclastic sands and muds. *Journal of Geology*, Vol.105, 173-191.

Numan, N.M.S. (1997). A Plate tectonic scenario for the Phanerozoic succession in Iraq. *Iraqi Geological Journal*, Vol. 30, No. 2, 85-110.

Pettijohn, F.J. Potter, P.E. & Siever, R. (1987). Sand and Sandstones. Springer, New York, 553 pp.

Ranjan, N. & Banerjee, D.M. (2009). Central Himalayan crystalline as the primary source for the sandstone-mudstone suites of the Siwalik Group: New geochemical evidence. *Gondwana Research*, Vol.16, No. 3-4, 687-696.

Roser, B.P. & Korsch, R.J. (1986). Determination of tectonic setting of sandstone-mudstone suites using SiO2 content and K_2O/Na_2O ratio. *Journal of Geology*, Vol. 94, 635-650. Bhatia M.R. (1985). Plate tectonics and geochemical composition of sandstones: a reply. *Journal of Geology*, Vol. 93, 85-87.

Roser, B.P. & Korsch, R.J. (1988). Provenance signatures of sandstone-mudstone suite determined using discriminant function analysis of major-element data. *Chemical Geology*, Vol. 67,119-139.

Sharland, P.R., Archer, R., Casey, D.M., Davies, R.B., Hall, S.H., Heward, A.P., Horbury, A.D. & Simmons, M.D. (2001). Arabian Plate Sequence Stratigarphy, *GeoArabia Special Publicatio 2*, Gulf Petrolink, Bahrain, 371p

Sloss, L.L. (1963). Sequences in the cratonic interior of North America, Geological Society of America Bulletin, Vol.74, 93-114.

Suttner, L.J. & Dutta, P.K. (1986). Alluvial sandstone composition and paleoclimate, 1. Framework mineralogy. *Journal of Sedimentary Petrology*, Vol. 56, No.2, p. 329-345.

Suttner, L.J., Basu, A. & Mack, G.H. (1981). Climate and origin of quartzarenites. *Journal of Sedimentary Petrology*, Vol.51, 1235-1246.

Sun, S. S. & McDonough, W. F. (1989). Chemical and isotopic systematic of oceanic basalt: implication for mantle composition and processes. In: Saunders, A. D. & Norry, M. J., *Magmatism in the oceanic basins*, Spec. Publ. Geol. Soc. London. Vol. 42, 313-346.

Taylor, S.R. & McLennan, S. (1985). The Continental Crust, its Composition and Evolution, Blackwell, Oxford, 312p.

Taylor, S.R. & McLennan, S.M. (1995). The geochemical evolution of the continental crust. *Reviews in Geophysics*, Vol. 33: 241-265

Umazano, A.M., Bellosi, E.S., Visconti, G., Jalfin, A.G. & Melchor, R.N. (2009). Sedimentary record of a Late Cretaceous volcanic arc in central Patagonia: petrography, geochemistry and provenance of fluvial volcaniclastic deposits of the Bajo Barreal Formation, San Jorge Basin, Argentina. *Cretaceous Research*, Vol.30, 749-766.

van Bellen, R.C., Dunnington, H., Wetzel, R. & Morton, D.M. (1959). Lexique Stratigraphique International, Paris, centre National recherché Scientifique Fasc. 10a, Iraq, 333pp.

Weltje, G.J. (1994). Provenance and dispersal of sand-sized sediments: reconstruction of dispersal patterns and sources of sand-size sediments by means of inverse modeling techniques, PhD thesis, Geologicaa Ultraiectina.

Wani, H. & Mondal, M.E.A. (2010). Petrological and geochemical evidence of the Paleoproterozoic and the Meso-Neoproterozoic sedimentary rocks of the Bastar craton, Indian Peninsula: Implications on paleoweathering and Proterozoic crustal evolution. *Journal of Asian Earth Sciences*, Vol.38, 220-232.

Yerino, L.N. & Maynard, J.B. (1984). Petrography of 7 modern marine sands from the Peru-ChileTrench and adjacent areas. *Sedimentlogy*, Vol.31, 83-89.

Zimmermann, U. & Spalletti, L.A. (2009). Provenance of the Lower Paleozoic Balcarce Formation (Tandilia System, Buenos Aires Province, Argentina): Implications for paleogeographic reconstructions of SW Gondwana. *Sedimentary Geology*, Vol.219, 7-23.

Zuffa, G.G. (1987). Unravelling hinterland and offshore paleogeography of deep-water arenites. In, J.K. Leggett and G.G. Zuffa (Eds.). Marine Clastic Sedimentology, Concepts and Case Studies, London, Graham and Trotman, 39-61.

8

Organic Petrology: An Overview

Suárez-Ruiz Isabel

Instituto Nacional del Carbón (INCAR-CSIC) Oviedo
Spain

1. Introduction

Organic Petrology is a branch of the Earth Science that studies fossil organic matter in sedimentary sequences including coal and the finely dispersed organic matter in rocks (DOM). It was developed from coal petrology that dates back to the end of the 19th century. The organic petrology is usually expressed by two fundamental parameters: the nature and proportions of the organic constituents, and by the rank or maturity of these organic components. In the case of coals the amount and composition of its mineral matter (grade of a coal) is another parameter to be taken into account. The work carried out by the International Committee for Coal and Organic Pertrology (ICCP) related to the development of the maceral nomenclature, classification, standardization and the use of petrographic methods has been reported in various editions of the " International Handbook of Coal Petrology" (ICCP, 1963, 1971, 1975, 1993) and other publications such as ICCP (1998, 2001) and Sýkorová et al. (2005). There are also several monographs focused on organic petrology such as those by Stach et al., (1982) and Taylor et al., (1998). Recently Suárez-Ruiz and Crelling (2008) edited a book on coal petrology and coal utilization. Other references in which organic petrology and/or coal petrology are discussed at some length include Ward (1984); Bustin et al. (1985); Falcon and Snyman (1986); Diessel (1992) and Teichmüller (1989). This chapter is an overview focused on organic petrology (including coal petrology), its fundamental concepts, the analytical techniques and the main current applications.

2. Fundamentals and general considerations

The particulate organic matter in sedimentary sequences ranges from disseminated occurrences of organic particles to concentrated organic matter in coals. The classification of the organic matter based on H/C and O/C atomic ratios (van Krevelen, 1993) distinguishes three main types of kerogens (Type I, Type II and Type III) that initially were associated to specific geological settings (Tissot and Welte, 1984). Vandenbroucke and Largeau (2007) revised this classification indicating that Type I kerogen may derive from various highly specific precursors characterized by a common high aliphaticity in different sedimentary environments, Type II kerogen can be associated with planktonic organic matter in open marine and fresh water lacustrine environments, and that Type III kerogen from higher plants can be associated with terrestrial inputs into lacustrine or marine settings. The kerogen classification (Fig. 1) is important because kerogen types (organic matter types) are directly related to total hydrocarbon potential, oil chemistry, and hydrocarbon generation kinetics. Coal, usually described as Type III kerogen, is a combustible sedimentary rock composed of lithified plant debris. This plant debris was originally deposited in a swampy depositional environment. The prolonged burial of the peat at depths of up to several kilometers, compaction, pressure and the influence of elevated temperatures for long periods of time (million years) are known as the coalification process that change peat into coal. Coal did not appear until the Devonian period due to the lack of terrestrial plants, although some organic matter derived from marine algae occurs in Precambrian sedimentary rocks.

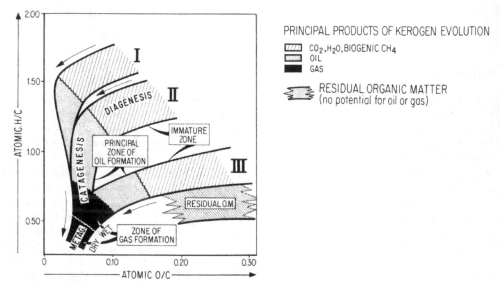

Fig. 1. Scheme of kerogen (organic matter) evolution on the van Krevelen's diagram. (Source: Petroleum Formation and Occurrence. 1984. By B.P. Tissot and D.H., Welte. 2nd ed., Springer-Verlag, 699 pp. This figure is from Chapter 7, Fig. II.7.2 in p. 216 (modified after Tissot, 1973). Copyright 1984, with kind permission from Springer Science and Business Media).

2.1 Origin, type and properties of organic components

Macerals are the microscopic organic components typically identified in coals. They derive from terrestrial, lacustrine and marine plant remains, and their appearance is a function of the parent material, of initial decomposition before and during the peat stages and also of the degree of evolution undergone. Macerals are distinguished from one to another on the basis of their physico-optical properties and universal acceptance is given to the ICCP classification of macerals (ICCP, 1963, 1971, 1975, 1998, 2001; Sýkorová et al., 2005) in three groups: liptinite, inertinite and huminite/vitrinite. These groups are subdivided into a variety of maceral sub-groups, macerals, and maceral varieties (Table 1).

The liptinite maceral group (Fig. 2) includes the optically distinct parts of plants such as spores, cuticles, suberine, etc., some degradation products, and those generated during the coalification/maturation process. Liptinite macerals show the highest content in hydrogen, contain compounds of mainly an aliphatic nature, and by thermal evolution they produce hydrocarbons (Tissot and Welte, 1984; Taylor et al., 1998; Wilkins and George, 2002). The inertinite maceral group derives from plant material that was strongly altered and degraded before deposition, or at the peat stage (Taylor et al., 1998; ICCP, 2001). Inertinite macerals (Fig. 2) exhibit a high degree of aromatization and condensation. The huminite/vitrinite maceral groups (Fig. 2) originate mainly from lignin and cellulose, partly also from tannins of the woody tissues of plants, and from colloidal humic gels (ICCP, 1971, 1998; Sýkorová et al., 2005). The formation of huminite/vitrinite macerals requires a set of successive processes such as humification, and biochemical and geochemical gelification. Huminite is only identified in low rank coals and is the precursor of the vitrinite macerals in medium and high rank coals. The chemical structure of the huminite/vitrinite is represented by aromatic and hydroaromatics compounds in low-rank coals, but with increasing coal rank the aromaticity, condensation and order of the polyaromatic units notably increase. Different maceral groups have different chemical and physical properties (Stach et al., 1982; van Krevelen, 1993; ICCP, 1998, 2001; Taylor et al., 1998; Sýkorová et al., 2005, and Suárez-Ruiz and Crelling, 2008).

Liptinite Group	Inertinite Group	Huminite Group		Vitrinite Group	
Sporinite	Fusinite	Telohuminite	Textinite	Telovitrinite	Telinite
Cutinite	Semifusinite		Ulminite		Collotelinite
Resinite	Funginite	Detrohuminite	Attrinite	Detrovitrinite	Vitrodetrinite
Alginite	Secretinite		Densinite		Collodetrinite
Suberinite	Macrinite	Gelohuminite	Corpohuminite	Gelovitrinite	Corpogelinite
Chlorophyllinite	Micrinite		Gelinite		Gelinite
Fluorinite	Inertoditrinite				
Bituminite					
Exudatinite					
Liptodetrinite					

Table 1. Main components of maceral groups. (Compiled from ICCP, 1971, 1975, 1998, 2001 and Sýkorová et al., 2005).

The three maceral groups are found in highest concentrations in sediments of terrestrial origin such as coals and carbonaceous shales, but they are nearly absent in most of carbonate rocks. In sedimentary rocks with DOM in addition to the described maceral groups, and secondary products, faunal relics and microfossils of various composition such as zooclasts, dinoflagellates and acritarchs can be found. Moreover, the non-structured organic matter or amorphous organic matter (Bertrand et al., 1993), intimately associated with the fine grain-minerals such as clays is also present in rocks with DOM. Another component of the DOM is the solid bitumen which appear in the macropores of the rocks, and as vein fillings. They are secondary products of the coalification/maturation processes and derive from the cracking of the macromolecular structure of organic matter into liquid hydrocarbons (of which they are the solid residue).

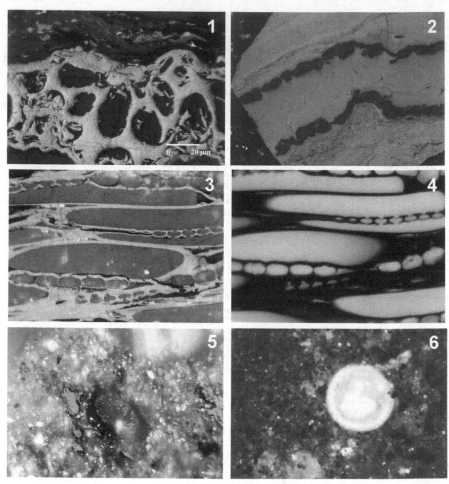

Fig. 2. Photomicrographs: 1,2,3,5: reflected white light; 4,6: fluorescence.
1,2) Bituminous coal with vitrinite, inertinite and liptinite. Carboniferous age, North Spain.
3,4) Telohuminite and resinite (liptinite) in a carbonaceous shale. Cretaceous, North Spain.
5,6) Oil shale (Tasmanite algae, Liptinite, right image). Jurassic, North Spain.

2.2 Mineral matter and microlithotypes in coals

Macerals in coals are accompanied by inorganic components that may appear as fine disseminations and as discrete partings. The inorganic fraction of coals should be taken into account due to its influence in coal utilization. Conventionally the mineral matter in coals has been differentiated on the basis of its origin into two major categories (ICCP, 1963; Stach et al., 1982; Bustin et al., 1985; Falcon and Snyman, 1986): i) the intrinsic inorganic matter which was present in the original plant tissues, and ii) the extrinsic or introduced forms of mineral matter that can be primary or secondary. The major groups of minerals in coals include clays, carbonates, iron sulphides and silicas. Other groups occur more rarely. Trace elements are also present in coals (Swaine ,1990; Finkelman, 1993).

The natural associations of macerals in coals are called microlithotypes. They are identified by optical microscopy and by definition (ICCP, 1963, 1971; Stach et al. 1982; Taylor et al., 1998) microlithotypes should form bands greater than 50 microns in width and should contain at least 5% of a maceral group. Microlithotypes are classified into three groups: mono-, bi-, and trimaceralic (or trimacerites) depending on whether their composition is made up of elements from one, two or three maceral groups. Macerals and microlithotypes that appear associated with minerals are named carbominerites.

2.3 Types of coal and lithotypes

The macroscopic composition of coals and their homogeneous / heterogeneous appearance is related to their composition. Macroscopically, all coals are classified in two categories or coal types: humic and sapropelic coals. Humic coals or banded coals are the most common in nature. Sapropelic coals are scarce and homogeneous in appearance. Lithotypes are the macroscopically recognizable bands in humic coals (ICCP, 1963) and four different lithotypes (vitrain, clarain, durain and fusain) have been described. Stach et al. (1982) extended the definition of lithotypes to include two varieties for sapropelic coals: cannnel and boghead coals.

2.4 Evolution of the organic matter

Coalification / maturation is a process that affects the organic matter after its deposition. This process is the result of the organic matter burial, and the corresponding increase in temperature and the time of this temperature influence. Pressure also at some stages of the coalification has some influence. Through the coalification process the original peat swamp is transformed, and passes through the progressive stages of evolution known as lignite, sub-bituminous, bituminous, anthracite and meta-anthracite (Fig. 3). With continued burial coalification may be followed by the graphitization process. The level of evolution reached by a coal through the coalification process is termed rank. The term coalification (Taylor et al., 1998) applies to coal evolution while the term maturation has been long used to describe the diagenetic evolution of the dispersed organic matter in sedimentary rocks leading to the formation of oil and gas (Tissot and Welte, 1984; Taylor et al., 1998; Vandenbroucke and Largeau, 2007).

As for the evolution (maturation) of disperse organic matter in sedimentary rocks, the process is parallel to the coalification of coals although the nomenclature of the various stages of maturity is different. Thus with increasing maturity of DOM in sedimentary rocks the following main stages were initially described (Tissot and Welte, 1984) and reviewed by Vandenbroucke and Largeau (2007): i) early diagenesis with a major loss of N and a potential incorporation of inorganic S and O into the organic matter, ii) diagenesis "sensu stricto" with a significant loss of O mainly as CO_2 and H_2O and it develops until the equivalent boundary in coals of sub-bituminous/bituminous rank stage; iii) catagenesis with a loss of H and C because the generation of oil and wet gas by thermal degradation of the organic matter due to the increase in temperature with burial in sedimentary basins. This stage follows the diagenesis and develops until the equivalent boundary in coals of bituminous/anthracite rank stage; and iv) metagenesis in which a reorganization of the aromatic network of the residual organic matter occurs increasing its aromaticity with production of CH_4 and non-hydrocarbon gases (CO_2, H_2S and N_2). It is the last stage of organic matter maturation and it is reached a great depths. With the increase of the organic evolution (diagenesis, catagenesis and metagenesis phases) the organic matter is progressively described as is immature, mature and overmature.

This information in combination with the amount and type of kerogen (organic matter) is of capital importance for oil and gas exploration (Fig. 3).

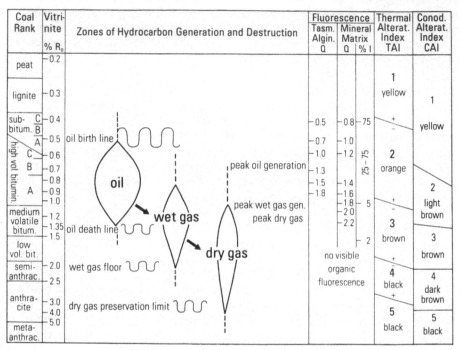

Fig. 3. Rank stages determined from microscopic rank/maturity parameters in relation to oil and gas production. (Source: Organic Petrology. 1998. By G.H. Taylor, M. Teichmüller, A. Davis, C.F.K. Diessel, R. Littke, P. Robert. Gebrüder Borntraeger. Berlin. 704 pp. This figure is from Chapter 3, Fig. 3.40 in p.135 (from Teichmüller 1987, after Dow 1977). Copyright 1998, with kind permission from Gebrüder Borntraeger. www.borntraeger-cramer.de).

3. Procedures in organic petrology

3.1 Sampling and preparation of samples for microscopic analysis

For petrographic studies of organic material, samples can be taken from the outcrops and open pit mines (surface samples) or from the underground mines, boreholes and exploration wells (subsurface samples). In some cases washed well cuttings may be used although special care is needed because cuttings can be contaminated by caved materials or drilling mud additives. In surface samples the main problem are weathering and oxidation processes that can lead to an extensive physical deterioration of the coals, particularly in case of low-rank coals. Adequate and precise guidelines for sampling procedures are given by international standards (ISO or ASTM norms).

The most widespread method of petrographic examination of coal and DOM in sedimentary rocks is the use of reflected white light microscopy on polished surfaces of high quality, under oil inmersion to enhance component or maceral reflectance differences. The polished surfaces can be obtained from polished epoxy-mounted samples which are prepared from whole rock, crushed material (coals or sedimentary rocks) or from concentrates of organic matter. The procedures of sample preparation for this type of analysis are standardized in ISO and ASTM norms.

3.2 Petrographic methods: Identification of organic matter

Petrographic analysis in incident light microscopy involves the identification of organic components, its quantification and rank/maturity determinations.

The identification of the organic matter types and the observation of all their physico-optical characteristics provides information about its source. Among these characteristics it should be mentioned (*e.g.,*) the organic and mineral matter associations; the shape and morphology of the organic components, intergrowths, organic distribution, porosity, internal reflections, texture and structures, fluorescence properties, optical isotropy/anisotropy, oxidation traces, etc. The optical texture (isotropic/anisotropic character) of the organic matter is a property to be considered in the case of organic components in rocks or in coals of high rank. This is especially significant in the case of anthracitic coals, bitumens in paragenesis with ore minerals, etc. With the coalification, the organic matter (initially isotropic and structurally disordered) becomes more ordered and tends to develop optical anisotropy.

3.3 Maceral and microlithotype analysis in coals

In geological research on coal basins, in the evaluation of coal seam quality or for coal utilization it is always necessary to know the quantitative composition of a coal in terms of macerals (and minerals in some cases) or maceral groups. This is because differences in maceral composition indicate differences in chemical composition and therefore in the technological properties of the coal. The maceral analysis in coals is performed using optical microscopy with a point-counter coupled to the microscope stage. The microscope stage is moved through the point-counter in a series of fixed intervals according to the total points to be recorded on the whole sample, and the identity of the maceral falling beneath the cross-hairs or micrometer after each advance is recorded. For that the microscope should be equipped with incident white light, oil immersion objectives (25x-50x magnification) and 8x to 10x oculars, one of which must contain an adjustable eyepiece with a micrometer or cross-hair. The analytical procedure for maceral analysis is standardized by the ISO and ASTM normative protocols. Automated methods using computerized image analysis were also developed in the last decades for maceral analysis although the manual point-counting method is the most extensively used. Although maceral analysis is carried out in white light, supplementary observations in fluorescence mode are recommended. Results of maceral analysis are reported on a volume percent basis.

Microlithotypes, and coal-mineral associations, carbominerites, can be also quantified via microlithotype analysis. The analytical procedures are standardized in the ISO norm. Microlithotype analysis is carried out in a similar way as maceral analysis. However, a suitable 20 points reticule (ICCP 1963; Stach et al. 1982) must be placed in one of the oculars in substitution of the micrometer or cross-hairs. In microlithotype analysis it is necessary to consider two conventions: i)- the minimum band width of the association to be measured must be 50 microns, and ii)- the 5% rule, which indicates that macerals present in the association in amounts smaller than 5% should be disregarded. Each observation on a 20-intersection or reticule is regarded as a point in the analysis. Under these circunstances each intersection of the reticule represents 5% of the total number of intersections (20) according to the 5% rule. For each set of readings at least ten reticule intersections must cover the coal particles. Microlithotype analysis may also include coal-mineral associations. Results from microlithotype analysis are also expressed as volume percent.

3.4 Quantification of organic component in organic-rich rocks other than coals

Quantification of organic components present in non-coal rock samples in reflected light microscopy and on particulate pellets can be also performed following the method of maceral analysis. However, taking into account that the organic matter in non-coal samples is disperse, and that many components are only visible in fluorescence mode results will not be as accurate and some components will be under- or over-estimated. To improve this type of analysis Boucsein and Stein (2009) used the point-counting method but only counting organic components versus minerals obtaining good results in their study of a black shale formation.

3.5 Optical methods to evaluate the coal rank and thermal maturity of DOM

The determination of thermal maturation of organic matter in coals and organic-rich rocks is essential to the geothermal studies in basin analysis, in the evaluation of natural fossil fuel resources, etc,.

3.5.1 Vitrinite reflectance measurements

The degree of evolution of the organic matter is commonly expressed in terms of vitrinite reflectance. Vitrinite reflectance is measured in incident white light microscopy and therefore, is an optical parameter that serves to describe the degree of coalification reached by a coal, the class of coals (ISO 11760, 2005), and the level of maturity of sedimentary rocks with DOM (ICCP, 1971; Stach et al., 1982; Taylor et al., 1998). The reflectance is a property related to the aromaticity of the organic components and it increases for all macerals as the level of coalification/maturation increases and the atomic O/C and H/C ratios decrease. However, reflectance usually is measured on vitrinite in coals and in rocks younger than Upper Silurian which marks the first appearance of vascular plants and so the precursor material of huminite and vitrinite. The procedure used for vitrinite reflectance measurements is standardized in ISO and ASTM norms for coals. Measurements of reflectance are carried out by comparing the amount of light reflected from a maceral (vitrinite) with the amount of light reflected by a standard and these measurements must be achieved in monochromatic green light (546 nm), in oil immersion and objectives with magnifications between 25x and 50 x. Two types of reflectance measurements can be made to quantify the reflectance of a vitrinite: i) the random reflectance (Ro, %), and ii) the maximum reflectance (Rv_{max}, %). Random reflectance is the reflectance of a particle in the orientation in which it is encountered (polarizer is removed from the microscope). For measurements of maximum reflectance the polarizer should be placed in the 45° position into the incident light beam. When the microscope stage is rotated through 360° the maximum reflectance can be taken (twice).

Measurements of vitrinite reflectance in sedimentary rocks with DOM follows a similar procedure as in coals. Vitrinite reflectance directly correlates with the other physico-chemical rank/maturity parameters and has been extensively and worldwide used to assess rock strata across a broad range of thermal maturity stages, ranging from early diagenesis through catagenesis to low grades of metamorphism in coals, oil shales, source rocks, etc.,

Anisotropy. Vitrinite in high rank coals and in rocks with overmature organic matter shows an anisotropic behavior and exhibits bi-reflectance. This is because with the increase in rank or maturity the structure of the carbonaceous material is reorganized and consequently almost all of its physical properties vary according to the considered section of the particle. With data of the true maximum and minimum reflectances, the anisotropy can be calculated as the difference: $R_{max}-R_{min}$. The anisotropy is linked to the overlying pressure of the strata and tectonic stress.

3.5.2 Reflectance measurements on zooclasts and solid bitumens

In absence of vitrinite in pre-Devonian rocks or when the sedimentary sequence is vitrinite-poor, the reflectance of zooclasts such as graptolites, chitinozoans, and scolecodonts have been used as rank/maturity parameter. Zooclast reflectance has been found to increase with the depth of burial and the increase in temperature. Graptolites possess weak to strong anisotropic character (Goodarzi and Norford, 1985) and its maximum reflectance also correlates with the Conodont Alteration Index (CAI) and optical properties of co-occurring coal macerals and solid bitumen. Other microfossils that have been used as maturation indicators are the chitinozoans and scolecodonts (Goodarzi and Higgins, 1987; Bertrand, 1990; Tricker et al., 1992). The reflectance of these microfossils increases at different rates depending on zooclast type.

Thermal evolution causes regular changes in the chemical composition of solid bitumens as well as an increase in its aromaticity and therefore, in the reflectance values. Several correlations have been developed between the reflectances of solid bitumen and vitrinite (Jacob, 1989, 1993; Bertrand, 1993; Landis and Castaño, 1995). However, the use of solid bitumen reflectance as a maturity parameter has been also a subject of some debate. For example, different relationships found between solid bitumen and vitrinite reflectances are thought to be due to the existence of various genetic types of solid bitumens with different optical properties. Despite of this, there are many studies that have used the reflectance properties of solid bitumens in conjunction with other thermal maturity parameters.

3.5.3 Fluorescence properties of organic constituents

Fluorescence microscopy is employed in coal petrology and in organic matter studies for characterization of liptinite macerals, organic matter composition, rank/maturation studies (as an essential criterion for oil and gas formation), hydrocarbon detection, and for correlation of technological properties of coals (thermoplastic, coking and oxidation features) to vitrinite fluorescence characteristics. Fluorescence analyses should be carried out using a reflected light microscope coupled to a photomultiplier (ICCP, 1975, 1993) o camera detector. The microscope must be equipped with a high pressure mercury or xenon lamp for illumination with the corresponding excitation filters to select the UV or blue light, barrier filters and also a variable interference filter covering the range of 400 to 700 nm. The fluorescence properties are primarily estimated on a qualitative basis (organic component diagnosis) using blue light excitation. Because the fluorescence of organic components is variable in intensity and color depending on the type and maturity level of the organic matter, these characteristics can be measured by monochromatic fluorescence microscopy and spectral fluorescence. The former includes quantitative measurements of fluorescence intensity at a specific wavelength (546 nm) recorded in relation to a standard (Jacob, 1980). Spectral fluorescence measurements determine changes in fluorescence color by recording the emission of the spectrum (Ottenjann, et al., 1975) between 400 - 700 nm. They also measure the spectral alteration or changes in fluorescence properties after 30 minutes of irradiation of organic substances (Teichmüller and Ottenjann, 1977). The classical reported parameters (Teichmüller, 1987) of fluorescence emission are: I= Fluorescence intensity at 546 nm; Lambda max = Spectral maximum; Q = Spectral quotient (red/green ratio); AI = Alteration of fluorescence intensity at 546 nm, and AS = Spectral alteration, in each case for a period of 30 minutes irradiation. Other spectral parameters are described by Teichmüller and Durand (1983) and Bertrand et al. (1993).

3.6 Other analysis in organic-rich rocks and coals
In transmitted light microscopy the organic matter is characterized by means of palynofacies analysis (Combaz, 1980; Tyson, 1993, 1995). The main populations identified in palynological residues were initially classified (Combaz, 1980) as: terrestrial organic fragments, pelagic and benthic microfossils and the so-called amorphous organic matter or amorphous fraction, and later reorganized as phytoclasts, palynomorphs and amorphous organic matter by Tyson (1993, 1995). The palynofacies analysis combined with studies of organic petrography, geochemical investigations, stratigraphy and paleontology, is a tool in the interdisciplinary analysis of organic matter providing accurate information in the investigation of organic facies and depositional paleoenvironments, paleoclimate reconstructions, origin and transfer studies of fossil organic matter in recent environments, hydrocarbon source rock potential and petroleum exploration.

In organic geochemistry one of the most commonly techniques to investigate DOM in rocks is the Rock-Eval pyrolysis (Espitalié and Bordenave, 1993 and references therein). Biomarkers obtained from gas chromatography and gas chromatography/mass spectrometry techniques have been found to be a powerful tool for source rocks to oil correlation (e.g., Philp, 1985a,b; Bordenave, 1993). The pyrolysis gas chromatography/mass spectrometry is another method to analyze the organic-rich rocks that provides more detailed information about the chemical composition of the organic components (e.g., Iglesias et al., 2002; Peters et al., 2005). Other bulk geochemical analyses used to characterize dispersed organic matter include total organic carbon determinations (TOC), elemental analysis (C, H, O, N, S), infrared spectroscopic analysis, nuclear magnetic resonance for characterizing the chemical structures and functional groups (e.g., Iglesias et al., 2006; Kelemen et al., 2007), electron paramagnetic resonance, and isotopic analyses (e.g., Galimov, 2006). Thermogravimetric analysis have been also used for characterization of organic matter in sedimentary rocks.

The conventional method to quantify the mineral matter in coals is the low temperature ashing that removes the organic matter by exposing the coal sample to a reactive oxygen plasma. The residue remaining (LTA) is evaluated by X-ray diffraction techniques (Renton, 1986). Mineral particles in coal can be also analyzed by scanning electron microscopy and similar techniques (e.g., Ward et al., 1996) such computer-controlled scanning electron microscopy (CCSEM) techniques (Gupta et al., 1999) and the advanced QEM*SEM and QEM Scan systems (Creelman and Ward, 1996) that allow the integration of SEM data with image analysis methods. The composition of specific minerals may be obtained from electron microprobe analysis using methods outlined by Reed (1996). Mössbauer spectroscopy, thermal analysis, Fourier-transform infra-red spectrometry, proton microprobe using proton-induced X-ray emission, the ion microprobe mass analyzer, the laser microprobe mass analyzer, laser ablation microprobe-inductively coupled plasma- mass spectrometry, the high-resolution ion microprobe, and X-ray absorption fine structure spectroscopy have also been used to identify the minerals, inorganic elements and trace elements in coals. To determine the concentration of individual inorganic elements in coal and in coals-ash (HTA, high temperature ashes) at major and trace levels (Huggins, 2002) several techniques are in use such as: X-ray fluorescence spectrometry, neutron activation analysis, atomic absorption spectrometry, optical emission spectrometry and inductively coupled plasma vaporization combined with atomic emission spectrometry or mass spectrometry.

3.7 Technological assays in coal utilization

The characterization of coals for industrial utilization requires specific analysis and tests (see Ward, 1984; Suárez-Ruiz and Crelling, 2008). Chemical analysis such as proximate analysis (moisture, ash, volatile matter, and fixed carbon contents), ultimate analysis (C, H, N, S, and O contents), and Calorific value (heating value) are conventional and routine determinations that are carried out in coal characterization. These analysis are normalized by the ISO and ASTM standards. Other coal quality parameters are obtained from specific assays such as: the Hardgrove grindability index (HGI) which indicates the ease by which the coal can be ground to fine powder; the ash fusion temperatures (AFT) that indicate the behavior of the ash residues from the coal at high temperatures; the free swelling index (FSI) or crucible swelling number (CSN) which measures the increase in volume of the coal when it is heated in a small crucible in the absence of air; the Roga index that provides information on the caking properties of the coal; the Gray-King and Fischer assays which determine the proportions of coke or char (solids), tar (organic liquids), liquor (ammonia-rich solutions) and gas produced when the coal is carbonized in the absence of air; and the Gieseler plastometer and Audibert-Arnu dilatomer tests that inform how the coal behaves in specific conditions. The Gieseler plastometer measures the coal's fluidity whereas the Audibert-Arnu dilatometer measures the contraction and coal's expansion.

4. Applications of organic petrology

The organic petrology studies throughout the identification and characterization of organic components, quantification, and assessment of rank/maturation stage of organic matter are mainly applied to geosciences investigations (*e.g.,* basin analysis), fossil fuel resources exploration and coal utilization. In the last years organic petrology has been also used in other fields such as environmental pollution and anthropogenic impacts, coal fires, archaeologic aspects and forensics.

4.1 Basin analysis

In the reconstruction of the palaeoenvironment of coal bearing sequences, in addition to the sedimentology and stratigraphy, the intrinsic characteristics of coals which are linked with the original vegetation and accumulation conditions should be considered. Petrographic composition of coal (macerals and microlithotypes) contributes to the definition of facies peat-forming for the palaeoenvironmental interpretation. On the basis of maceral composition of coals Diessel (1986); Mukhopadhyay (1986); Calder et al. (1991); Kalkreuth et al. (1991); Hacquebard (1993), and Jasper et al. (2010) have proposed a series of petrographic indices. By correlation of these indices the authors have defined facies and palaeoenvironmental diagrams that permit the assessment of the depositional environment and paleoconditions at the time of peat formation such as the vegetation type, preservation, the relative water level (relative dryness or wetness), the rates of peat accumulation and the basin subsidence. Many authors have used those petrographic indices in studies of coal-bearing sequences such as (e.g.,) Diessel (1992); Jiménez et al. (1999); Piedad-Sánchez et al. (2004); Kalaitzidis et al. (2010). In the definition of coal facies and for palaeoenvironmental interpretations other authors have used the microlithotypes associations and lithotypes (e.g., Hacquebard and Donaldson, 1969; Marchioni and Kalkreuth, 1991).

Taking into account the relationship between the degree of coalification/maturation and the rock temperature to which the organic matter has been heated during its geological history, many attempts have been carried out to reconstruct paleogeothermicity using vitrinite reflectance as the rank/maturity parameter. Sweeney and Burnham (1990) early indicated the need of using kinetic calculations for the evolution of vitrinite reflectance rather than simple vitrinite reflectance gradients in order to obtain reliable geothermal data because the measures of rank/maturity cannot be directly converted into paleotemperatures values. These authors proposed the kinetic EASY%R algorithm that permits the comparison of measured and calculated vitrinite reflectance values and thus the effective calibration of the thermal and burial histories. This model has been used (e.g.,) by Hertle and Littke (2000); Petersen et al. (2009). Carr (2003) noticed that in overpressure systems a pressure-dependent kinetic model should be used for modeling. Models based on heat flow but also including data on compaction, pressure, temperature, maturation of organic matter, petroleum generation, migration and accumulation through time have been used to simulate the burial history of sediments by Tissot and Welte (1984); Baur et al. (2010), among others. However, in tectonically active basins, the resulting maturation patterns are the sum of a wide set of processes that should be taken into account in basin modeling.

Organic facies, palaeoenvironment interpretations and rank/maturation data are necessary for the reconstruction of geological history of sedimentary basins. Rank profiles and rank maps ilustrating the regional rank variations (e.g., Suárez-Ruiz and Prado, 1995; Colmenero et al., 2008; Ruppert et al., 2010) provide important information in basin studies and modeling. Detailed research focused on the palaeogeography and/or the palaeothermal conditions during the geological evolution of basins all over the world have been reported (e.g.,) by Büker et al. (1996); Baur et al. (2010).

4.2 Fossil fuel resources

For fossil fuel resources exploration the knowledge of all the characteristics of the sedimentary basins is a pre-requisite. These natural resources are related with coal, source rocks, petroleum (oil and gas), reservoir rocks including coal as reservoir, oil shales, shale gas, black shales, carbonaceous shales, and coal bed methane. The quantification of their reserves is approached through exploration activities which include the creation of geological maps of the areas containing the natural resource (geological settings), the organic facies and sedimentary paleoenvironment investigations, geophysical surveys, and finally the exploration drilling. The last step is the construction of a geological model of the basin which provides with an accurate report of the area to be developed and reduce the exploratory risks.

4.2.1 Source rocks

The source rocks are organic-rich sediments that can be originated in various sedimentary environments such as deep marine, lacustrine and deltaic paleoenvironments. According to Law (1999) source rocks can be divided into four major categories: i) potential source rocks, that include rocks containing organic matter in sufficient quantity to generate and expel hydrocarbons if subjected to an increase in thermal maturation, ii) effective rocks with organic matter that are generating and/or expelling hydrocarbons forming commercial accumulations, iii) relic effective rocks, which describes an effective source rock that ceased the generation and expulsion of hydrocarbons because a thermal cooling event such as uplift or erosion before exhausting its organic matter, and iv) spent rock which describes an exhausted source rock due to the lack of sufficient organic matter or because the rock is in an overmature state. The main information to be obtained in the characterization of potential source rocks is: i) amount of organic matter in the rock, ii) quality and type of organic matter capable of yielding hydrocarbons (different types of organic matter have different hydrocarbon potentials), and iii) maturity of organic matter. Examples can be found in (*e.g.,*) Tissot and Welte (1984); Bordenave (1993); Suárez-Ruiz and Prado (1995); Hakimi et al. (2010).

Tissot and Welte (1984) described oil shales as any immature rock that yields oil in commercial amounts upon pyrolysis. Oil shales are generated in a wide range of sedimentary environments, from terrestrial swamps and pools, transitional zones between land and sea and to deep marine basins. The organic matter is frequently marine or freshwater algae, but other planktonic organisms and also bacterial biomass may contribute significantly. Oil shales are widely distributed around the world and hundreds of deposits are known (*e.g.,* Kruge and Suárez-Ruiz, 1991; Fu et al., 2009).

Black shale is a dark-colored mudrock containing more or less transformed organic matter and silt- and clay-size mineral grains that accumulated together. The content of organic matter rarely exceeds a few percentages. If the content in organic matter is higher, the rock may yield petroleum by pyrolysis and then it is described as oil shale. In these shales the organic matter is made up of structureless and carbonized organic components, in which terrestrially-derived fragments may be present together with microplankton. Black shales can be originated in different depositional paleoenvironments. However, the most appropriate conditions include anoxic bottom waters with low energy, in which the sulphate-reducing bacteria generate hydrogen sulphide after oxygen has been used up, and the benthic life is inhibited (Hallam, 1980). Black shales are in some cases source rocks of metals but its major interest is that they may be important source for petroleum (*e.g.,* Kolonic et al., 2002). If they have generated liquid and/or gaseous hydrocarbons these products can be found in reservoir rocks. Recently the potential of black shale itself as a reservoir for gas hydrocarbons has been recognized.

Carbonaceous shales are a transition from humic coals into coaly shale. Taylor et al. (1998) published a detailed microscopic classification of organic matter-rich sediments designating the organic matter-rich rocks as carbonaceous when they contained terrigenous organic matter. There is an economic importance attached to these carbonaceous rocks, because they may be an important source of natural gas or methane (*e.g.,* Lee et al., 2010).

4.2.2 Reservoir rocks

Reservoir rocks are one of the elements of the petroleum system. Any permeable and porous rock may act as a reservoir for hydrocarbon and may be detrital or clastic rocks or precipitated rocks such as carbonates. Occasionally shale, volcanic rocks, and fractured basement can act as reservoir rocks. The characterization of a reservoir rock is fundamental for all kind of studies related to hydrocarbon field exploration. The two most essential elements of a reservoir rock are porosity and permeability. The majority of petroleum accumulations are found in clastic reservoir rocks being sandstones the most common. However, more than 40% of the so-called giant oil and gas fields are found in carbonates (Tissot and Welte, 1984). The study of reservoir rocks includes the analyses of secondary organic matter products (solid bitumens), and gaseous hydrocarbons which in the case of coals are called coalbed methane (Ayers, 2002). Coal is classified as a continuous-type, unconventional natural gas reservoir, with complex reservoir properties in which coalbed methane is stored dominantly in an adsorbed state rather than a free state. In coal reservoir studies diverse geologic factors influence storage capacity, hydrocarbon content, and production performance. The variability of those geologic factors is conditioning the specific strategies for both coalbed methane development and carbon sequestration in each basin. Coalbed methane (CBM) has diverse origins, namely thermogenic and biogenic and isotopic analysis helps in its differentiation (Whiticar, 1996). On CBM there is a large amount of papers (*e.g.*, Gentzis et al., 2006; Alsaab et al., 2008)

Shale gas is an unconventional natural-gas reservoir, and refers to in situ hydrocarbon gas present in fine grained and organic rich sedimentary rocks. Gas produced from organic-rich shales is of both biogenic and thermogenic origin and is stored in situ, in shales as both adsorbed gas (on organic matter) and free gas (in fractures or pores). The issues to be taken into account when characterizing shale gas are: the organic richness, its type or quality, and the thermal maturity which may be approached via organic petrology. Organic geochemistry and isotopic composition of gas are also a need for determining the hydrocarbon type and composition. Finally, generation and retention of hydrocarbons in the system are assessed via hydrocarbon modeling.

4.3 Coal utilization

Coal is another fossil fuel resource. Their physico-chemical properties are related to three independent parameters: rank, type and grade (Ward, 1984; Taylor et al., 1998). Coal quality is a function of these factors. The role of coal petrology in coal utilization has been recently compiled by Suárez-Ruiz and Crelling (2008).

4.3.1 Coal Mining and coal preparation

Coal composition and rank influence its behavior during mining and beneficiation. During mining coal is liberated from the host clastic rock and through beneficiation process a higher grade coal is obtained. The coal quality can be assessed by understanding the distribution of coal lithotypes relative to the thickness and splitting characteristics of a coal seam which is related to the original depositional controls. Microscopically, changes in maceral composition can also help in predicting splitting in advance of mining (Esterle and Ferm, 1986). Abrupt changes in coal rank or trace element content may indicate intrusive bodies and dykes and in some cases faults by a change in vitrinite reflectance. Development of anisotropy may also record changes in palaeo stress.

The analysis of the distribution of coal lithotypes, coal rank and grade may serve to predict the behavior of a coal seam at all stages of the mining and beneficiation chain. A significant hazard during mining, transport and storage of coal is the self-heating of coal that led to spontaneous combustion. Lower rank coals are generally more prone to self-heating than higher rank coals, but this is not a linear relationship and coal type, coal particle size and climate are important factors that influence this process. Coals with high contents in reactive macerals, (liptinites and vitrinites) are more prone to self-heating than inertinite- rich macerals. Beamish and Arisoy (2007) also demonstrated the importance of mineral type and content for self-heating.

4.3.2 Coal combustion

In coal combustion, coal rank is fundamental because it influences the heating value of a coal, and the combustion characteristics. The maceral composition is also fundamental to the combustion properties, as different maceral groups combust at different temperatures and rates. The inorganic composition is also basic to the heating rate of the coal. Mineral matter type and contents influence the ash yield, as well as the emission of gaseous oxides and trace elements in the combustion process. Neavel (1981) described a series of coal properties that are important in combustion such as: calorific value, grindability, combustibility (combustion properties of macerals), and ash properties. An important issue in coal combustion is the production of fly ash (Fig. 4). The characteristics of these fly ash will depend on the type and grade of combusted coals and the combustion conditions. Fly ash can be re-used and methods used in organic petrology serves to control the composition and properties of fly ash (Suárez-Ruiz-and Crelling, 2008).

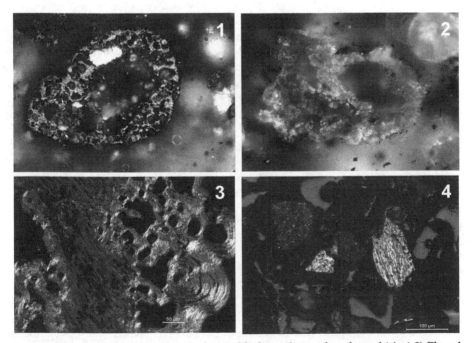

Fig. 4. Images in optical microscopy. Polarized light and retarder plate of 1 λ. 1,2) Fly ash from pulverized coal combustion of bituminous coals; left: unburned carbon, right: inorganic particles. 3) Metallurgical coke. 4) Gasification residue of a subbituminous coal.

4.3.3 Coal carbonization

Coal carbonization is the process for producing metallurgical coke (Fig. 4) for use in iron-making blast furnaces and other metal smelting processes. Coal carbonization is developed when coal is heated to temperatures as high as 1100°C in the absence of oxygen. In addition to coke, the by-products obtained in this process are tars, light oils, coke oven gas along with ammonia, water, and sulfur compounds that are also thermally removed from the coal. Coal petrology has been applied to general aspects of carbonization such as to evaluate the suitability of a given coal for coking, in the prediction of the strength of cokes made from both single coals and coal blends, the peak coking pressures, and the coke reactivity. Coke shows distinct optical textures and structures and these features can be investigated by petrographic analysis to understand the behavior of coke. There are classification systems of the coke structures published in the literature. Coal petrology has been also incorporated into models for prediction of coke strength of single coals and coal blends. In coke making the peak coking pressure is a factor to be considered, and Benedict and Thompson (1976) developed petrographic methods to predict peak coking pressure.

4.3.4 Coal gasification

Gasification is a conversion process that is described as the reaction of solid fuels with air, oxygen, steam, carbon dioxide, or a mixture of these gases at temperatures exceeding 800ºC in a reducing environment where the air/oxygen ratio is controlled. The obtained gaseous products are further processed for use as an energy source or as a raw material. The chemical composition of the gas produced depends on coal composition and rank, coal preparation; gasification agents, gasification conditions, and plant configuration. The performance of the gasification process is also dependent on the type of the coal and the gasifier configuration. Elemental composition (organic and inorganic) is one of the relevant coal properties together with surface characteristics and porosity, and intrinsic reactivity (van Heek and Muhlen, 1986). Coal rank affects hydrogen and oxygen ratios, gasifier performance and the char morphology (Harris et al., 2006). Char morphology can be predicted from the initial maceral and microlithotype composition because they behave in specific ways when exposed to increasing temperatures and oxygen-rich or oxygen-depleted environments. The analysis of the type and amount of mineral matter in coals permits the prediction of the ash or slag properties, the conversion rate, the size of the ash bed in the reactor, and ash / slag handling facilities following conversion.

4.3.5 Direct coal liquefaction

Coal liquefaction can be carried out by different processes (directly by hydrogenation, indirectly by the Fischer-Tropsch synthesis, and removing some carbon from coal by pyrolysis) which differ in their requirements, technology, yield of fuel and final products. The objective of direct coal liquefaction (conversion of coal to liquids by hydrogenation) is to add hydrogen to the organic structure of the coal, breaking it down to produce distillable liquids to be used as transportation fuels or chemicals. A series of coal characteristics should be controlled such as coal rank and composition which are primary factors that influence the liquefaction behavior (e.g., Cudmore 1984). The maceral composition of the coals has also a strong influence on the composition of the resulting products (liquid yield), and the mineral matter content of the coal influence the liquefaction behavior. The coal liquefaction process also produces solid residues and their study provide information on the physical and chemical transformation of coal macerals and minerals in the process. Moreover the study of solid residues by petrography also informs about the efficiency of the conversion process. In 1993 the ICCP published a Classification of Hydrogenation Residues in order to standardize the terminology and organize the residue components by their degree of reaction or mode of formation.

4.3.6 Coal-derived carbon materials

Coal and its by-products may also be employed as a source of organic chemicals and carbon based-materials (Schobert and Song, 2002) such as carbon fibers and carbon-carbon composites, graphites, activated carbons, and carbon foams among others. Carbon materials are composed of a high percentage of carbon and taking into account that coals by definition are carbon-rich solids a wide range of solid carbon materials can be obtained from coal and its by-products. Depending on the rank and composition, coals may be used directly as precursors to obtain solid carbon materials. Solid residues with a high carbon content from coal utilization are also investigated as precursors of carbon materials. In the case of coal by-products, coal-tar is produced during coal conversion to obtain metallurgical coke. Pitches obtained from coal-tar are in turn used as precursors for chemicals and carbon materials. In addition, mesophase pitches (the anisotropic fraction of pitches) obtained from coal-tar pitches, are also used as precursors of carbon materials. In the study of the physico-optical properties of solid carbon materials and their precursors, the procedures used in organic petrography play a major role. This is because the relevant parameters obtained from microscopic analysis of a material or precursor can be correlated with other analytical information assisting in the interpretation of the properties and behaviour of a particular carbon material. The most relevant characteristics analysed by the microscopic examination of carbon materials and their precursors are reported inSuárez-Ruiz and Crelling (2008).

4.4 Ore deposits

It is known that some ore deposits occur associated to the organic matter in variable amounts and that some of them are of economic interest. The involvement of the organic matter in some aspects of the ore formation (*e.g.,* Meyers et al., 1992; Mossmann, 1999) varies from active participation in the emplacement of ore deposits to post-depositional alteration of the organic matter that may be related or unrelated with the ore forming process. The incorporation of organic petrology methods in studies of metal concentration may assist in the evaluation of the time-temperature burial history of ores and may elucidate the active or passive role of the organic matter in ore deposition, thereby helping to determine the source of metals, transportation and precipitation mechanisms, and provides indications of the thermal history of the host rocks. The microscopic approach is the same as that conventionally applied to coal/ coke and for the dispersed organic matter in sedimentary rocks. Some examples of this type of studies include (*e.g.,*) Parnell et al. (1993); Parnell (2001) and Glikson and Mastalerz (2000).

4.5 Other fields of organic petrology application

In the last years organic petrology studies also complements investigations developed in other different fields not strictly related to those previously described such as coal fires, environmental impacts, archaeology and provenance of organic artifacts, and forensics.

Coal fires in un-mined outcrops, abandoned mines and coal waste piles constitute a serious safety and environmental hazard. There is an increasing interest in the self-heating processes (*e.g.*), Stracher and Taylor (2004). Critical coal properties requisite for spontaneous combustion are summarized in Suárez-Ruiz and Crelling (2008) and include: high moisture and volatile matter contents; particle size and available surface area which permits the permeation of air and water; mineral matter type; petrographic composition (presence of reactive macerals), and coal rank.

Organic petrology has been also used in combination with other analytical techniques in investigations of environmental pollution due to anthropogenic activities. Some examples include those by Cohen et al. (1999a,b) and Carrie et al. (2009). The identification of organic particulates derived from industrial activities such as coal mining, preparation, transport, blending, storage and utilization, coke, coal-tar, pitchs, etc., is important due to the existing relationships between such organic particulates, their sorption properties, and the presence of PCBs, PAHs, and PCDD/Fs organic pollutants (*e.g.*, Yang et al., 2008). Taking into account this Crelling et al. (2006) published an atlas focused on organic particles from anthropogenic activities that can be identified by using petrographic methods.

The possibilities of the organic petrology to investigate the archaeological objects of organic nature such as jet have been revised by Teichmuller (1992) and recently by Suárez- Ruiz and Crelling (2008). Most of the scientific research into the occurrences of jet from different countries, its nature, origin, properties and quality has been carried out in the last three decades (*e.g.*, Petrova et al., 1985; Suárez-Ruiz et al., 1994 among others). The organic petrography in combination with organic geochemistry has demonstrated that jet is a perhydrous coal with suppressed reflectance, high H/C atomic ratio and high oil yields. Because the scarcity of this material, jet ornaments are very appreciated and organic petrology helps to differentiate jet from other apparently similar materials.

Petrologic methods have been also used to investigate the provenance of organic objects and artifacts from archeological and similar sites (*e.g.*, Smith, 2005). Moreover, the organic petrography also helped in the assessment of trading patterns of coals during the past two centuries in relation to (*e.g.*,) the shipwrecks (Erskine et al., 2008).

Forensic geology is mainly concerned with studies of rocks, sediments, minerals, soils and dusts and it can be defined as the discipline that uses geological methods and materials in the analysis of samples and places that maybe connected with criminal behavior or disasters (Murray, 2004; Ruffell, 2010). Therefore forensic geology includes geological methods of analysis such as geophysics, petrography, geochemistry, microscopy and micropaleontology (Ruffell, 2010).

5. Summary

This paper provides a general picture of organic petrology and the significant role it plays in contributing to scientific knowledge. Organic petrology is a subject of broad scope that started out as coal petrology for coal research (coal exploration and utilization). This discipline dates back to the end of the 19th century. The two basic concepts of the organic petrology, composition and rank/maturation of organic matter and the undeniable economic interest shown in fossil fuels have extended its application to the field of hydrocarbon resources. In the last few decades, due to its high versatility, the organic petrology has been increasingly branching out into other fields of studies such as coal fires, environmental impacts, archaeology and provenance of organic artifacts, and forensics. Consequently a great deal of research has been carried out in both, fundamental and applied organic petrology as it is reflected in hundreds of published papers, and monographs.

6. References

Alsaab, D., Elie, M., Izart, A., Sachsenhofer, R.F., Privalov, V.A., Suarez-Ruiz, I., Martinez, L., 2008. Comparison of hydrocarbon gases (C1-C5) production from Carboniferous Donets (Ukraine) and Cretaceous Sabinas (Mexico) coals. International Journal of Coal Geology 74, 154-162.

Ayers, W.B., Jr., 2002. Coalbed gas systems, resources, and production and a review of contrasting cases from the San Juan and Powder River Basins: AAPG Bulletin 86, 1853-1890.

Baur, F., Littke, R., Wielens, H., Lampe, C., Fuchs, T., 2010. Basin modeling meets rift analysis – A numerical modeling study from the Jeanne d'Arc basin, offshore Newfoundland, Canada. Marine and Petroleum Geology 27, 585-599.

Beamish, B.B., Arisoy, A., 2007. Effect of mineral matter on coal self-heating rate. Fuel 87, 125–130.

Benedict, L.G., Thompson, R.R., 1976. Selection of coals and coal mines to avoid excessive coking pressure. A.I.M. E. Ironmaking Proc. 35, 276-288.

Bertrand, R., 1990. Correlations among the reflectances of vitrinite, chitinozoans, graptolites and scolecodonts. Organic Geochemistry 15, 565-574.

Bertrand, R., 1993. Standardization of solid bitumen reflectance to vitrinite in some Paleozoic sequences of Canada. Energy Sources 15, 269-287.

Bertrand, Ph., Bordenave, M.L., Brosse, E., Espitalié, J., Houzay, J.P., Pradier, B., Vandenbroucke, M., Walgenwitz, F., 1993. Other methods and tools for source rock appraisal. In: Bordenave, M.L., (Ed.). Applied Petroleum Geochemistry. Ed. Technip. Paris. Chapter II. 3, 279-371.

Bordenave, M.J. (Ed.) 1993. Applied Petroleum Geochemistry. Éditions Technip, Paris. 352 pp.

Boucsein, B., Stein, R., 2009. Black shale formation in the late Paleocene/early Eocene Arctic Ocean and paleoenvironmental conditions: New results from a detailed organic petrological study. Marine and Petroleum Geology 26, 416–426.

Büker, C., Littke, R., Welte, D.H., 1996. 2D-modeling of the thermal evolution of Carboniferous Devonian sedimentary rocks of the eastern Ruhr basin and northern Rhenish Massif, Germany. Z. Dt. Geol. Ges. 2, 146.

Bustin, R.M., Cameron, A.R., Grieve, D.A., Kalkreuth, W.D., 1985. Coal Petrology, its principles, methods and applications. Geological Association of Canada, Short Course Notes, 2nd edition, Victoria, British Columbia 3, 230 pp.

Calder, J.H., Gibling, M.R., Mukhopadhyay, P., 1991. Peat formation in a Westphalian B piedmont setting, Cumberland basin, Nova Scotia: implications for the maceral-based interpretation of rheotrophic and raised paleomires. Bulletin de la Societe Geologique de France 162, 283-298.

Carr, A.D., 2003. Thermal history model for the South Central Graben, North Sea, derived using both tectonics and maturation. International Journal of Coal Geology 54, 3–19.

Carrie, J., Sanei, H., Goodarzi, F., Stern, G., Wang, F., 2009. Characterization of organic matter in surface sediments of the Mackenzie River Basin, Canada. International Journal of Coal Geology 77, 416-423.

Cohen, A.D., Gage, C.P., Moore, W.S., 1999a. Combining organic petrography and palynology to assess anthropogenic impacts on peatlands Part 1. An example from the northern Everglades of Florida. International Journal of Coal Geology 39, 3–45.

Cohen, A.D., Gage, C.P., Moore, W.S., VanPelt, R-S. 1999b. Combining organic petrography and palynology to assess anthropogenic impacts on peatlands Part 2. An example from a Carolina Bay wetland at the Savannah River Site in South Carolina. International Journal of Coal Geology 39, 47-95.

Colmenero, J.R., Suárez-Ruiz, I., Fernández-Suárez, J., Barba, P., Llorens, T., 2008. Genesis and rank distribution of Upper Carboniferous coal basins in the Cantabrian Mountains, Northern Spain. International Journal of Coal Geology 76, 187–204.

Combaz, A., 1980. Les kérogenes vus au microscope. In: Durand, B., (Ed.). Kerogen. Insoluble organic matter from sedimentary rocks. Paris. 55-113.

Creelman, R.A., Ward, C.R., 1996. A scanning electron microscope method for automated, quantitative analysis of mineral matter in coal. International Journal of Coal Geology 30, 249-269.

Crelling, J., Glickson, M., Huggett, W., Borrego, M.A.G., Hower, J., Ligouis, B., Mastalerz, M., Misz, M., Suárez-Ruiz, I., Valentim, B., 2006. Atlas of anthropogenic particles. Mastalerz, M., Hower, J.C., (Eds.), International Committee for Coal and Organic Petrology and Indiana Geological Survey, CD-ROM.

Cudmore, J.F., 1984. Coal utilization. In: Ward, C.R., (Ed.). Coal Geology and Coal Technology. Blackwell Scientific Publications, Melbourne, Chapter 4, 113-150.

Diessel, C.F.K., 1986. On the correlation between coal facies and depositional environments. Symposium Advance in the Study of the Sydney Basin. Proceedings, 20th, Newcastle, 1986, 19-22.

Diessel, C.F.K., 1992. Coal-Bearing Depositional Systems. Springer-Verlag, Berlin, 721 pp.

Erskine, N., Smith, A.H.V., Crosdale, P.J., 2008. Provenance of Coals Recovered from the 33 Wreck of HMAV Bounty. The International Journal of Nautical Archaeology 37/1, 34 171-176.

Espitalié, J., Bordenave, M.L., 1993. Source rock parameters. In: Bordenave,35 M.L., (Ed.), Applied petroleum geochemistry 2, 237-272.

Esterle, J.S., Ferm, J.C., 1986. Relationship between petrographic and chemical properties and coal seam geometry, Hance seam, Breathitt Formation, southeastern Kentucky. International Journal of Coal Geology 6, 199-214.

Falcon, R.M.S., Snyman, C.P., 1986. An introduction to coal petrography: Atlas of petrographic constituents in the bituminous coals of Southern Africa. The Geological Society of South Africa. Review Paper, 2. Kelvin House, 2 Hollard Street. (Johannesburg).

Finkelman, R.B., 1993. Trace and minor elements in coal. In: Engel, M.H., Macko, S.A., (Eds.), Organic Geochemistry. Plenum Press. New York. 593-607.

Fu, X., Wang, J., Zeng, Y., Li, Z., Wang, Z., 2009. Geochemical and palynological investigation of the Shengli River marine oil shale (China): Implications for paleoenvironment and paleoclimate. International Journal of Coal Geology 78, 217-224.

Galimov, E.M., 2006. Isotope organic geochemistry. Organic Geochemistry 37, 1200-1262.

Gentzis, T., Schoderbek, D., Pollock, S., 2006. Evaluating the coalbed methane potential of the Gething coals in NE British Columbia, Canada: An example from the Highhat area, Peace River coalfield. International Journal of Coal Geology 68, 135-150.

Glickson, M., Mastalerz, M., (Eds.), 2000. Organic Matter and Mineralisation: Thermal Alteration, Hydrocarbon Generation and Role in Metallogenesis. Kluwer Academic Publishers, Great Britain, 454 pp.

Goodarzi, F., Norford, B.S., 1985. Graptolites as indicators of the temperature histories of rocks. Journal of the Geological Society 142, part 6, 1089-1099. London.

Goodarzi, F., Higgins, A.C., 1987. Optical properties of scolecodonts and their use as indicators of thermal maturity. Marine Petroleum Geology 4, 353-359.

Gupta, R.P., Yan, L., Kennedy, E.M., Wall, T.F., Masson, M., Kerrison, K., 1999. System accuracy for CCSEM analysis of minerals in coal. In: Gupta, R.P., Wall, T.F., Baxter, L., (Eds.), Impact of Mineral Impurities in Solid Fuel Combustion. Kluwer Academic/ Plenum Publishers, New York, 225-235.

Hacquebard, P.A., Donaldson, J.R., 1969. Carboniferous Coal Deposition Associated with Flood-Plain and Limnic Environments in Nova Scotia. In: Dapples, E.D., Hopkins, M.E., (Eds.), Environments of Coal Deposition. Geological Society of America. Boulder, Colo. (Special Paper 114), 143-191.

Hacquebard, P., 1993. The Sydney coalfield of Nova Scotia, Canada. International Journal of Coal Geology 23, 29–42.

Hakimi, M.H., Abdullah, W.H., Shalaby, M.R., 2010. Source rock characterization and oil generating potential of the Jurassic Madbi Formation, onshore East Shabowah oilfields, Republic of Yemen. Organic Geochemistry 41, 513-521.

Hallam, A., 1980. Black Shales. J. geol. Soc. London 137, 123-124.

Harris, D.J., Roberts, D.G., Henderson, D.G., 2006. Gasification behaviour of Australian coals at high temperature and pressure. Fuel 85/2, 134 – 142.

Hertle, M., Littke, R., 2000. Coalification pattern and thermal modelling of the Permo-Carboniferous Saar Basin, SW-Germany. International Journal of Coal Geology 42, 273–296.

Huggins, F.E., 2002. Overview of analytical methods for inorganic constituents in coal. International Journal of Coal Geology 50, 169-214.

Iglesias, M. J., del Río, J. C., Laggoun-Défarge, F., Cuesta, M. J., Suárez-Ruiz, I., 2002. Control of the chemical structure in perhydrous coals by FTIR and Py-GC/MS. Journal of Analytical and Applied Pyrolysis 62/1, 1-34.

Iglesias, M.J., Cuesta, M.J., Laggoun-Défarge, F., Suárez-Ruiz, I., 2006. 1D-NMR and 2D-NMR analysis of the thermal degradation products from vitrinites in relation to their natural hydrogen enrichment. Journal of Analytical and Applied Pyrolysis 77, 83-93.

International Committee for Coal Petrology (ICCP). 1963. International Handbook of Coal Petrography. 2nd Ed. Centre National de la Recherche Scientifique. Academy of Sciences of the USSR. Paris, Moscow.

International Committee for Coal Petrology, (ICCP), 1971. International Handbook of Coal Petrography, 1st Supplement to 2nd Edition. CNRS (Paris).

International Committee for Coal Petrology, (ICCP), 1975. International Handbook of Coal Petrography, 2nd Supplement to 2nd Edition. CNRS (Paris).

International Committee for Coal Petrology, (ICCP), 1993. International Handbook of Coal Petrography, 3rd Supplement to 2nd Edition. University of Newcastle on Tyne (England).

International Committee for Coal and Organic Petrology, (ICCP), 1998. The new vitrinite classification (ICCP System 1994). Fuel 77, 349–358.

International Committee for Coal and Organic Petrology (ICCP), 2001. The new inertinite classification (ICCP System 1994). Fuel 80, 459–471.

ISO 11760, 2005. Classification of coals. International Organization for Standardization. 1st edition, Geneva, Switzerland. 9 pp.

Jacob, H., 1980. Die Anwendung der Mikrophotometrie in der organischen Petrologie. Leitz. Mitt. Wiss Techn., 7, 209-216.

Jacob, H., 1989. Classification, structure, genesis and practical importance of natural solid oil bitumen ("migrabitumen"). International Journal of Coal geology 11, 65-79.

Jacob, H., 1993. Nomenclature, Classification, Characterization, and Genesis of Natural Solid Bitumen (Migrabitumen). In: Parnell, J., Kucha, H., Landais, P., (Eds.). Bitumens in ore deposits. Special Publication of the Society for Geology Applied to Mineral Deposits 9. Springer-Verlag, 11-27

Jasper, K., Hartkopf-Fröder, C., Flajs, G., Littke R., 2010. Evolution of Pennsylvanian (Late Carboniferous) peat swamps of the Ruhr Basin, Germany: Comparison of palynological, coal petrographical and organic geochemical data. International Journal of Coal Geology 83, 346–365.

Jiménez, A., Martínez Tarazona, M.R., Suárez-Ruiz, I., 1999. Paleoenvironmental conditions of Puertollano coals (Spain): petrological and geochemical study. International Journal of Coal Geology 41, 189– 211.

Kalaitzidis, K., Siavalas, G., Skarpelis, N., Araujo, C.V., Christanis, K., 2010. Late Cretaceous coal overlying karstic bauxite deposits in the Parnassus-Ghiona Unit, Central Greece: Coal characteristics and depositional environment. International Journal of Coal Geology 81, 211–226.

Kalkreuth, W., Marchioni, D., Calder, J., Lamberson, M., Naylor, R., Paul, J., 1991. The relationship between coal petrography and depositional environment from selected coal basins in Canada, International Journal Coal Geology 19, 21–76.

Kelemen, S.R., Afeworki, M., Gorbaty, M.L., Sansone, M., Kwiatek, P.J., Walters, C.C., Freund, H., Siskin, M., 2007. Direct characterization of kerogen by X-ray and solid-state 13C nuclear magnetic resonance methods. Energy & Fuels 21, 1548-1561.

Kolonic, S., Sinninghe Damsté, J.S., Böttcher, M.E., Kuypers, M.M.M., Kuhnt, W., Beckmann, B., Scheeder, G., Wagner, T., 2002. Geochemical characterization of Cenomanian/Turonian black shales from the Tarfaya Basin (SW Morocco). Journal of Petroleum Geology 25/3, 325-350.

Kruge, M., Suárez –Ruiz, I., 1991. Organic geochemistry and petrography of Spanish oil shales. Fuel 70, 1298-1302.

Landis, Ch., Castaño, J., 1995. Maturation and bulk chemical properties of a suite of solid hydrocarbons. Organic Geochemistry 271, 137-149.

Law, C. A. 1999. Evaluating Source Rocks. In: AAPG Special Volumes. Volume Treatise of Petroleum Geology/Handbook of Petroleum Geology: Exploring for Oil and Gas Traps, Pages 3-1 - 3-34.

Lee, H.-T. , Tsai, L.L. , Sun, L.-C., 2010. Relationships among geochemical indices of coal and carbonaceous materials and implication for hydrocarbon potential evaluation. Environmental Earth Sciences 60/3, 559-572.

Marchioni, D., Kalkreuth, W., 1991. Coal facies interpretations based on lithotype and maceral variations in Lower Cretaceous (Gates Formation) coals of Western Canada. International Journal of Coal Geology 18, 125–162.

Meyers, P.A., Pratt, L.M., Nagy, B., 1992. Introduction to geochemistry of metalliferous black shales. Chemical Geology 99, vii-xi.

Mossman, D.J., 1999. Carbonaceous substances in mineral deposits: implications for geochimemical exploration. Journal of Geochemical Exploration 66, 241-247.

Mukhopadhyay, P. K., 1986. Petrography of selected Wilcox and Jackson Group lignites from the Tertiary of Texas. In: Finkelman, R.B., Casagrande, D.J., (Eds.). Geology of Golf Coast Lignites, Field Trip Guide Book. Geological Society of America. Boulder, Colo, p. 140.

Murray, R. C., 2004. Evidence form the earth. Mountain Press Publishing Co., Missoula, Montana, 226 pp.

Neavel, R.C., 1981. Origin, petrography and classification of coal. In: Elliott, M.A., (Ed.), Chemistry of Coal Utilization. 2nd Supplementary Volume. New York, John Wiley and Sons, 91- 158.

Ottenjann, K., Teichmuller, M, Wolf, M., 1975. Spectral fluorescence measurements of sporinites in reflected light and their applicability for coalification studies. In: Alpern, B., (Ed.). Pétrographie de la matière organique des sédiments, relations avec la paléotempérature et le potentiel pétrolier. CNRS, Paris. 67-91.

Parnell, J., Kucha, H., Landais, P., (Eds.), 1993. Bitumens in ore deposits. Special Publication No 9 on the Society for Geology Applied to Mineral Deposits. Springer-Verlag, 520 pp.

Parnell, J., 2001. Paragenesis of mineralization within fractured pebbles in Witwatersrand conglomerates. Mineralium Deposita 36, 689-699.

Peters, K.E., Walters, C.C., Moldowan, J.M., 2005. The biomarker guide volume 2: biomarkers and isotopes in petroleum exploration and earth history. Cambridge University Press, Cambridge, 475-1155.

Petersen, H.I., Sherwood, N., Mathiesen, A., Fyhn, M.B.W., Dau, N.T., Russell, N., Bojesen-Koefoed, J.A., Nielsen, L.H., 2009. Application of integrated vitrinite reflectance and FAMM analyses for thermal maturity assessment of the northeastern Malay Basin, offshore Vietnam: Implications for petroleum prospectivity evaluation. Marine and Petroleum Geology 26, 319-332.

Petrova, R., Mincev, D., Nikolov, Z.D.R., 1985. Comparative investigations on gagate and vitrain from the Balkan Coal Basin. International Journal of Coal Geology 5, 275-280.

Philp, R.P., 1985a. Biological markers in fossil fuel production. Mass Spectrometry reviews 4, 1-54.Philp, R.P., 1985b. Fossil fuel biomarkers. Applications and Spectra. Methods in Geochemistry and Geophysics 23. Elsevier, Amsterdam. 294 pp.

Piedad-Sánchez, N., Suárez-Ruiz, I., Martínez, L., Izart, A., Elie, M., Keravis, D., 2004. Organic petrology and geochemistry of the Carboniferous coal seams from the Central Asturian Coal Basin (NW Spain). International Journal of Coal Geology 57, 211-242.

Reed, S.J.B., 1996. Electron Microprobe Analysis and Scanning Electron Microscopy in Geology. Cambridge University Press, Cambridge, 201 pp.

Renton, J.J., 1986. Semiquantitative determination of coal minerals by X-ray diffractometry. In: Vorres, K.S., (Ed.,). Mineral Matter and Ash in Coal. American Chemical Society Symposium Series 301, 53-60.

Ruffell, A., 2010. Forensic pedology, forensic geology, forensic geoscience, geoforensics and soil forensics. Forensic Science International 202, 9-12.

Ruppert, L.F. Hower, J.C., Ryder, R.T. Levine, J.R., Trippi, M.H., Grady, W.C. 2010. Geologic controls on thermal maturity patterns in Pennsylvanian coal-bearing rocks in the Appalachian basin. International Journal of Coal Geology 81, 169-181.

Schobert, H.H., Song, C., 2002. Chemicals and materials from coal in the 21st century. Fuel 81, 15-32.

Smith, A.H.V., 2005. Coal microscopy in the service of archaeology. International Journal of Coal Geology 62, 49-59.

Stach, E., Mackowsky, M-Th., Teichmuller, M., Taylor, G.H., Chandra, D., Teichmuller, R., (Eds.)., 1982. Coal Petrology. Gebruder Borntraeger (Berlin - Stuttgart), 535 pp.

Stracher, G.B., Taylor, T.P., 2004. Coal fires burning out of control around the world: thermodynamic recipe for environmental catastrophe. International Journal of Coal Geology 59, 7-17

Suárez-Ruiz, I., Iglesias, M.J., Jiménez Bautista, A., Laggoun-Defarge, F., Prado, J.G., 1994. Petrographic and geochemical anomalies detected in the Spanish Jurassic jet. In: Mukhopadhyay, P.K., Dow, W.G., (Eds.). Vitrinite reflectance as a maturity parameter. Applications and limitations. American Chemical Society Symposium Series. ACS Books 570, 76-92.

Suárez-Ruiz, I., Prado, J. G., 1995. Characterization of Jurassic black shales from Asturias (Northern Spain): Evolution and petroleum potential. In: Snape, C., (Ed.), Composition, Geochemistry and Conversion of Oil Shales, NATO A.S.I. Series, Series C: Mathematical and Physical Sciences 455, 387-395.

Suárez-Ruiz, I., Crelling, J.C., (Eds.), 2008. Applied Coal petrology. The role of petrology in coal utilization. Elsevier, Amsterdam, 398 pp.

Swaine, D.J., 1990. Trace Elements in Coal. Butterworths, London, 278 pp.

Sweeney, J.J., Burnham, A.K., 1990. Evaluation of a simple model of vitrinite reflectance based on chemical kinetics. The American Association of Petroleum Geologists Bulletin 74, 1559-1570.

Sýkorová, I., Pickel, W., Christianis, K., Wolf, M., Taylor, G.H., Flores, D., 2005. Classification of huminite - ICCP System 1994. International Journal of Coal Geology 62, 85-106.

Taylor, G.H., Teichmuller, M., Davis, A., Diessel, C.F.K., Littke, R., Robert, P., 1998. Organic petrology. Gebrüder Borntraeger. Berlin. 704 pp.

Teichmüller, M., Ottenjann, K., 1977. Art und Diagenese von Liptiniten und lipoiden Stoffen in einem Erdolmuttergestein aufgrund fluoreszenzmikroskopischer Untersuchungen, Erdol, Kohle, Petrochem. 30, 387-398.

Teichmüller, M., Durand, B., 1983. Fluorecence microscopical rank studies on liptinite and vitrinite in peat and coals and comparison with results of the Rock-Eval pyrolysis. International Journal of Coal Geology 2, 197-230.

Teichmüller, M., 1987. Recent advances in coalification studies in their applications to geology. In: Scott, A.C., (Ed.). Coal and coal-bearing strata: Recent Advances geol. Soc. Spec. Publ., Blackwell Sci. Publ., Oxford. 32, 127-170.

Teichmüller, M., 1989. The genesis of coal from the wiewpoint of coal petrology. International Journal of Coal Geology 12, 1-87.

Teichmüller, M., 1992. Organic petrology in the service of the archaeology. International Journal of Coal Geology 20, 1-21.

Tissot, B.P., Welte, D.H., 1984. Petroleum Formation and Occurrence. 2nd Edition. Berlin, Springer-Verlag, 699 pp.

Tricker, P.M., Marshal, J.E.A., Badman, T.D., 1992. Chitinozoan reflectance: a Lower Palaeozoic thermal maturity indicator. Marine and Petroleum Geology 9, 302-307.

Tyson, R.V., 1993. Palynofacies analysis. In: Jenkings, D.G., (Ed.), Applied micropaleontology Kluwer Academic Publishers, Dordrecht. The Netherlands, 153-191.

Tyson, R.V., 1995. Sedimentary organic matter: Organic facies and palynofacies. Chapman and Hall, London. 615 pp.

van Heek K.H, Muhlen H.-J., 1986. Effect of coal and char properties on gasification. In: Proceedings of the first International Rolduc Symposium on Coal Science, Rolduc, (Moulijn, J.A., Kapteijn F., Eds.), 113-133.

van Krevelen, D.W., 1993. Coal: Typology - Chemistry -Physics - Constitution, 3rd ed. Elsevier, The Netherlands. 979 pp.

Vandenbroucke, M., Largeau, C., 2007. Kerogen origin, evolution and structure. Organic Geochemistry 38, 719-833.

Ward, C.R. (Ed.), 1984. Coal Geology and Coal Technology. Blackwell Scientific Publications, Melbourne, 345 pp.

Ward, C.R., Corcoran, J.F., Saxby, J.D., Read, H.W., 1996. Occurrence of phosphorus minerals in Australian coal seams. International Journal of Coal Geology 31, 185-210.

Whiticar, M.J., 1996. Stable isotope geochemistry of coals, humic kerogens and related natural gases. International Journal of Coal Geology 32, 191-215.

Wilkins, R.W.T., George, S.C., 2002. Coal as a source rock for oil: a review. International Journal of Coal Geology 50, 317– 361.

Yang, Y., Ligouis, B., Pies, C., Grathwohl, P., Hofmann, T., 2008. Occurrence of coal and coal-derived particle-bound polycyclic aromatic hydrocarbons (PAHs) in a river floodplain soil. Environmental Pollution 151, 121-129.

Permissions

The contributors of this book come from diverse backgrounds, making this book a truly international effort. This book will bring forth new frontiers with its revolutionizing research information and detailed analysis of the nascent developments around the world.

We would like to thank Ali Al-Juboury and John Shervais, for lending their expertise to make the book truly unique. They have played a crucial role in the development of this book. Without their invaluable contribution this book wouldn't have been possible. They have made vital efforts to compile up to date information on the varied aspects of this subject to make this book a valuable addition to the collection of many professionals and students.

This book was conceptualized with the vision of imparting up-to-date information and advanced data in this field. To ensure the same, a matchless editorial board was set up. Every individual on the board went through rigorous rounds of assessment to prove their worth. After which they invested a large part of their time researching and compiling the most relevant data for our readers. Conferences and sessions were held from time to time between the editorial board and the contributing authors to present the data in the most comprehensible form. The editorial team has worked tirelessly to provide valuable and valid information to help people across the globe.

Every chapter published in this book has been scrutinized by our experts. Their significance has been extensively debated. The topics covered herein carry significant findings which will fuel the growth of the discipline. They may even be implemented as practical applications or may be referred to as a beginning point for another development. Chapters in this book were first published by InTech; hereby published with permission under the Creative Commons Attribution License or equivalent.

The editorial board has been involved in producing this book since its inception. They have spent rigorous hours researching and exploring the diverse topics which have resulted in the successful publishing of this book. They have passed on their knowledge of decades through this book. To expedite this challenging task, the publisher supported the team at every step. A small team of assistant editors was also appointed to further simplify the editing procedure and attain best results for the readers.

Our editorial team has been hand-picked from every corner of the world. Their multi-ethnicity adds dynamic inputs to the discussions which result in innovative outcomes. These outcomes are then further discussed with the researchers and contributors who give their valuable feedback and opinion regarding the same. The feedback is then collaborated with the researches and they are edited in a comprehensive manner to aid the understanding of the subject.

Apart from the editorial board, the designing team has also invested a significant amount of their time in understanding the subject and creating the most relevant covers. They scrutinized every image to scout for the most suitable representation of the subject and create an appropriate cover for the book.

The publishing team has been involved in this book since its early stages. They were actively engaged in every process, be it collecting the data, connecting with the contributors or procuring relevant information. The team has been an ardent support to the editorial, designing and production team. Their endless efforts to recruit the best for this project, has resulted in the accomplishment of this book. They are a veteran in the field of academics and their pool of knowledge is as vast as their experience in printing. Their expertise and guidance has proved useful at every step. Their uncompromising quality standards have made this book an exceptional effort. Their encouragement from time to time has been an inspiration for everyone.

The publisher and the editorial board hope that this book will prove to be a valuable piece of knowledge for researchers, students, practitioners and scholars across the globe.

List of Contributors

Yan-Jie Tang, Hong-Fu Zhang and Ji-Feng Ying
State Key Laboratory of Lithospheric Evolution, Institute of Geology and Geophysics, Chinese Academy of Sciences, Beijing, China

Masao Ban and Naoyoshi Iwata
Department of Earth and Environmental Sciences, Yamagata University, Japan

Shiho Hirotani and Osamu Ishizuka
Geological Survey of Japan/AIST, Japan

Prosper M. Nude
Department of Earth Science, University of Ghana, P.O. Box LG 58, Legon-Accra, Ghana

Kodjopa Attoh
Department of Earth & Atmospheric Sciences, Cornell University, Ithaca, NY 14853, USA

John W. Shervais
Department of Geology, Utah State University, Logan UT 84322, USA

Gordon Foli
Department of Earth and Environmental sciences, University for Development studies, Navrongo Campus, Ghana

Hemayat Jamali
Geological Survey of Iran, Tehran, Iran
Tarbiat Moallem University, Tehran, Iran

Abdolmajid Yaghubpur, Behzad Mehrabi and Ahmad Meshkani
Tarbiat Moallem University, Tehran, Iran

Yildirim Dilek
Department of Geology, Miami University, Oxford, OH, USA

Farahnaz Daliran
Institute for Applied Geosciences, University of Karlsruhe, Karlsruhe, Germany

Blazo Boev
Faculty of Natural and Technical Science, "Goce Delcev" University – Stip, Republic of Macedonia

Rade Jelenkovic
Faculty of Mining and Geology, Belgrade University, Republic of Serbia

Delia del Pilar Montecinos de Almeida
Universidade Federal do Pampa (UNIPAMPA), Brazil

Farid Chemale Jr.
Universidade de Brasília (UnB), Brazil

Adriane Machado
Centro de Geofísica da Universidade de Coimbra (CGUC), Portugal

A. I. Al-Juboury
Research Center for Dams and Water Resources, Mosul University, Mosul, Iraq

Suárez-Ruiz Isabel
Instituto Nacional del Carbón (INCAR-CSIC) Oviedo, Spain

Printed in the USA
CPSIA information can be obtained
at www.ICGtesting.com
JSHW011421221024
72173JS00004B/630